To M...
Ha...
Lots of Love
Gabi + Rosie

The Heart of Victoria – Bendigo, Castlemaine, Daylesford, Macedon
The Spirit of Place

Published
November 2004

Published by
Best Shot! Publications Pty Ltd, ACN 100 252 926
2/16 Coronation St, Geelong West, Victoria 3218, Australia

ISBN 0-9756023-0-6

Text © Best Shot! 2004
Photos © as indicated Best Shot!, Gary Chapman, Daryl Chibnall, Greg 'Arjuna' Govinda, Theodore Halacas, Geoff Hocking, Martin Hurley, Joseph Kinsela, Chris Kirwan, Joe Mortelliti, Alison Pouliot, Christine Ramsay, Sandy Scheltema, Katherine Seppings, Pete Walsh

Cover: Pete Walsh, Daylesford
Inside front cover: Gary Chapman

Printed in China by Twin Age Limited, Hong Kong, email: twinage@netvigator.com

All rights reserved. No part of this publication may be reproduced, stored in a retrieval system or transmitted in any form by any means, electronic, mechanical, photocopying, recording or otherwise, except brief extracts for the purposes of review, without the written permission of the publisher and the copyright owner.

The Heart of Victoria
Bendigo, Castlemaine, Daylesford, Macedon

Joseph Kinsela

The Spirit of Place

BestShot!

Best Shot! Publications: Ballarat, Bendigo, Croydon, Geelong, Flinders, Taradale

CONTENTS

INTRODUCTION	7
GEOLOGY	13
ABORIGINAL / KOORI	15
EXPLORERS	17
Major Mitchell	17
Bourke & Wills Expedition	23
GOLD!	27
Alluvial Mining	27
Deep Lead Mining	29
DEMOCRACY	31
ENGINEERING	33
GARDENS	35
EVENTS & FESTIVALS	37
WALKING	41
The Great Dividing Trail	43
PICNIC SPOTS	45
ACTIVITIES	47
Heritage Transport	47
Horseriding	47
Shopping	47
Nurseries	49
Mineral Springs	49
Spas	49
PLACES TO STAY	51
FOOD	53
Eateries	53
Local Produce	53
WINE	57
THE ARTS	65
Bendigo Art Gallery	65
Castlemaine Art Gallery	67
Visual Artists	69
Galleries	71
Performing Arts	73
MUSEUMS	73
VISITOR INFORMATION CENTRES	77
MOUNT MACEDON	**79**
MACEDON	85
WOODEND	87
HANGING ROCK	89
LANCEFIELD & ROMSEY	91
KYNETON	93
Piper St	95
Mollison & High Sts	97
AROUND KYNETON	99
MALMSBURY	105
TARADALE	109
CASTLEMAINE	**111**
Castlemaine Gothic	119
Christ Church & Agitation Hill	121
Campbells Creek	121
CHEWTON	123
HARCOURT	125
MALDON	129
Mt Tarrangower	129
The Township	131
BENDIGO	**133**
Pall Mall Trams	137
Vienna-in-the-bush	139
Vahland & Getzschmann	141
Buildings & Precincts	143
WOODEND TO TRENTHAM	157
TRENTHAM	157
TRENTHAM TO BLACKWOOD	159
BLACKWOOD	161
NEWSTEAD	163
GUILDFORD	165
GUILDFORD TO CHEWTON	165
NEWSTEAD TO DAYLESFORD	167
MOUNT FRANKLIN	167
DAYLESFORD	**169**
Daylesford Botanic Gardens	169
Wombat Hill	169
The Convent Gallery	171
Howe & Vincent Sts	171
Spa Country Railway	171
HEPBURN SPRINGS	173

Opposite: Katherine Seppings, Expedition Pass

INTRODUCTION

The Heart of Victoria is not found in Melbourne but lies north-west of the city and the bay. Its natural signpost is the distant blue slope of Mount Macedon, seen from the Bolte and Westgate bridges. Sixty km along the Calder Highway, as the road is winding through old volcanic hills and the steep slope of the mountain looms up ahead, something has changed – the bare plains and the city smog lie behind, ahead is the real Victoria of great natural forests, myriad grassy hills and the fascinating old towns of the goldfields.

This land tells many tales. The earliest are of the ancient silted lakes of Gondwanaland, which rose up to form the tablelands of central Victoria. Over time, granite flows thrust through weak fault lines and stand now as bony ridges after millions of years of erosion have exposed them to the skies. Volcanic cones poured out lava flows to reshape the shallow valleys during the last 60 million years and all the while small rivers were cutting relentlessly through the surface of the land to form valleys and gorges.

Streams from the southern face of these ridges flow to the shores of Port Phillip Bay. Over the Dividing Range (the heights between Gisborne and Woodend) the flow is northward across the great plain to join the Murray River. Running down through the hills over millions of years, these creeks and rivers exposed seams of gold in the quartz ridges and carried precious flakes along the valley floors to lie as bait for human hordes in the mid-19th century.

It is hard to explain the manic phenomenon of gold fever, but it brought many thousands of people, men mainly, halfway around the globe in the hope of making their fortune. Imagine the puzzlement of Aboriginal eyes as their homelands were swamped with strangers clutching shovels, picks and wash pans. The pastoralists who had arrived only in the 1830s and 1840s must have been almost as astonished, as tent-towns sprang up on their pastureland.

Tents soon gave way to slab-and-bark huts, which soon gave way to brick and stone. Within ten years a double-track of railway line was pushing its way through the countryside from Melbourne. It was the most complete piece of rail engineering attempted in the colony, with solid bluestone tunnels and viaducts and the mighty Taradale bridge forty metres high. And it brought the world in comfort to Bendigo.

Opposite: Joe Mortelliti, Shepherd's Flat

Although Woodend marks the tableland of the Great Divide and Kyneton grew to be the solid agricultural and commercial heart of the new northwest en route to Mount Alexander Diggings, the gold itself – and the vital activity and human tumult – was to be found in the gullies of the downlands, Taradale, Chewton, Castlemaine, Maldon and most of all Bendigo.

Bendigo is the pot of gold at the end of the ranges. The name itself conjures up images of fabulous wealth and the great endeavours of mining engineers and entrepreneurs. Today this lively city is dominated by its splendid historical inheritance: domes and spires and porticoes, trams still moving along the principal boulevard, a vast gothic cathedral, and a wondrous collection of old pubs. A new mining venture, starting in 2005, plans to go deeper even than the deepest shafts under the city – which in their time were the deepest in the world – so Bendigo is once again dipping into the pot of gold.

The Heart of Victoria covers twelve old towns and cities and a couple of dozen villages. Today they are the fascinating places where atmosphere can be savoured – and the Australian experience is a close encounter.

Some of the wonderful and curious things to be found include:
- mineral springs with a well-developed spa and health resort
- at least five botanic gardens, one of which is on top of an ancient volcano
- two railway preservation projects with running trains
- two major regional art galleries, one with an international collection and the other entirely Australian
- many splendid private gardens of the 19th and 20th centuries
- restaurants and cafés with local and regional character
- more than 100 wineries in three regions: Macedon Ranges cool climate wines identified as light-bodied reds and delicate whites, the Harcourt region, and the Bendigo region
- orchards and berry-farms, and other gastronomic delights
- theatrical and musical presentations in the major centres, together with regular cultural festivals.

Above: Katherine Seppings, Bendigo
Opposite: Katherine Seppings, Moonlight Flat

THE HEART OF VICTORIA

BENDIGO, BLACKWOOD, CAMPBELLS CREEK, CASTLEMAINE, CHEWTON, DAYLESFORD,
GUILDFORD, HARCOURT, HEPBURN SPRINGS, KYNETON, LANCEFIELD, MACEDON,
MALDON, MALMSBURY, MT MACEDON, NEWSTEAD, ROMSEY, TARADALE, TRENTHAM

Opposite: Joe Mortelliti, Clydesdale

GEOLOGY

Most of the landscape of Victoria has been shaped by the action of volcanoes. Because they are not massive cone-shaped peaks like Mount Fuji or some of New Zealand's mountains, this volcanic activity is not often recognised.

On the Calder Freeway toward Mount Macedon each hill the traveller sees, almost without exception from the tallest peak to the slightest mound, has been an active volcano. The oldest, like Macedon itself, go back hundreds of million years, others like the little scoria mounds near Diggers Rest are quite recent.

Once past the Dividing Range near Woodend the pattern continues, and much of the level ground is lava flow that has filled ancient stream-beds. As this happened the river was displaced to find a new course, cutting down into softer rock beside the newly hardened flows.

Some waterfalls in the region actually fall over the edge of lava deposits that blocked up their old watercourses a few million years ago. The rivers kept running down their newly shallow courses because of higher ground to the sides, and are now wearing away the basalt faces at the end of the lava flow. Examples are seen at Trentham and Turpins Falls.

Geology manages geography, and no region illustrates this more clearly. The size of trees varies according to soils and history. Solid old redgums and yellowbox trees line the creeks, while stringybarks cover the scourged hillsides where miners sluiced away the topsoil, looking for the gold deposited millions of years earlier.

BENDIGO ART GALLERY

Established in 1887, Bendigo Art Gallery has undergone a major renovation program resulting in the restoration of the 19th century rooms and addition of new gallery spaces for sculpture and contemporary Australian art. Bendigo Art Gallery has an extensive and varied collection with an emphasis on 19th century European art and Australian art from 1800s onwards, and a dynamic temporary exhibition program.

Open 10am - 5pm daily (except Christmas Day) Free guided tours at 2pm
Tel: 03 5443 4991   **Email:** bendigoartgallery@bendigo.vic.gov.au
www.bendigoartgallery.com.au   Photograph © John Gollings 2003

Opposite: Joe Mortelliti, Bayton, Sedonia Ranges

ABORIGINAL / KOORI

Archaeologists continue to make significant discoveries that impact on our understanding of the human settlement of Australia, but most agree Aboriginals arrived at least 60,000 years ago. In Victoria stone tools and bones in the Keilor area of the Maribyrnong valley date back 30,000 years.

Over that time Kooris have seen volcanic eruptions come and go in several places, at least two 'ice ages', and rising sea-water form the inland sea we now call Port Phillip Bay. Geologists have found a stone axe head trapped in a basalt rock from a lava flow, and descriptions of volcanic activity are enshrined in tribal stories.

In the Lake Condah region of Western Victoria the Gunditj Mara people constructed channels, weirs and stone fish-traps. These allowed the Gunditj Mara to create ideal conditions for harvesting eels and fish, which in turn resulted in high population densities and the construction of permanent stone huts clustered into villages.

The people of the Kulin nation (or, more properly, linguistic group) that surrounded Port Phillip and Westernport Bays also built sophisticated fish traps. Among other finely crafted artifacts, the Kulin were notable for the elaborate possum-skin capes they made.

The Kooris of the Macedon Ranges were Kulin people, and part of the Wurrundjeri group. On the northern slopes of the Dividing Range, Kooris were members of the Dja Dja Wurrung. Their home country was the entire region now encompassed by the shires of Hepburn Springs and Mount Alexander, and Bendigo.

Reliable population figures do not exist, but it is likely there were never large numbers: perhaps 18,000 to 20,000 Kooris across Victoria. What is certain is that the moment Europeans arrived Koori numbers began to decline rapidly. Disease probably had a greater impact than violence, but there was certainly violence.

Large-scale Koori resistance to the European invasion of Victoria was centred around Lake Condah, and became known as the Eumeralla War (after the Eumeralla River). However, across the state, numerous skirmishes, mass poisonings and shootings continued for several decades. The fundamental issue was the failure of Europeans to understand Koori notions of territory and land occupation. Many clashes were sparked by the fact that Kooris did not distinguish between domesticated stock that in European eyes was 'owned' and therefore couldn't be hunted, and native fauna which was not 'owned' and therefore could be hunted.

Above: Katherine Seppings, Mt Tarrangower from Barkers Creek
Opposite: Katherine Seppings, Moonlight Flat

EXPLORERS

Australia's great explorers were the astronauts of the 19th century, and their paths across the Heart of Victoria can be traced through memorial statuary, stone cairns, names of places and buildings that pepper the region.

The excitement and hero-worship they generated can only be explained in the context of their own time. It is hard to comprehend, from our position in the 21st century, how utterly frustrated the early settlers were by their inability to find answers about the geography, resources and potential of Australia. For example, it took a full 25 years after European settlement at Sydney to pick out a route over the Blue Mountains, which began a mere 60 km west of the town. The continent did not give up its secrets easily and the great inland regions continued to defy party after party of explorers.

MAJOR MITCHELL

Major Sir Thomas Livingstone Mitchell, "Major Mitchell", Surveyor General to His Majesty's Colony of New South Wales, was one of the few explorers who did have success, and he is consequently the most famous. He explored the fertile pastures of western and northern Victoria in 1836.

Born in 1792, Mitchell entered the army at sixteen, making rapid progress until being appointed to the rank of Major in 1826 after being an aide to the Duke of Wellington. In 1827 he came to Sydney to take the position of Assistant-Surveyor to John Oxley. The following year he succeeded to Oxley's position through the latter's ill-health. He made several expeditions in his new capacity, chiefly tracing the course of the western rivers of NSW. Governor Bourke, disappointed at failure to resolve the riddle of the rivers that flowed inland (it was assumed they flowed into an inland sea) and that no fertile inland region had been discovered, sent Mitchell on a quest to explore the course of the Darling River and its connection with the Murray River.

Above: Sandy Scheltema, Wombat State Forest
Opposite: Pete Walsh, Wombat State Forest

MITCHELL'S 1836 EXPEDITION

Major Mitchell set out in March 1836 with a party of 24 armed men, and an ox-wagon carrying a boat, to explore the Darling and Murray rivers. Included in the party were two Aboriginal guides, known to the Europeans as John Piper and Tommy Came Last. By 3 June they had found the swirling junction of the Darling with the Murray, and the party turned upstream along the Murray, reaching Swan Hill at the end of the month. It was a wet winter, and Mitchell was excited by the quality of the soils and agricultural potential of the land.

Leaving the Murray near Kerang the party went southwest and encountered the dramatic wall of mountains, which were named the Grampians after the Scottish ranges they do indeed resemble. Mitchell continued south-west via the Wannon and Glenelg rivers toward Portland where he encountered the Henty family, who had settled the area from Launceston the previous year and established pastoral and whaling enterprises.

On their return journey, the party crossed ranges which Mitchell named the Pyrenees. The name must have resulted from the delusionary effect of having been so long on the plains: the border ranges of France and Spain reach 3000 metres; Victoria's Pyrenees may just reach 1000.

Mitchell continued to be impressed by the regular streams and extensive grassland. When the party reached the Loddon River, the bullock cart bogged and it took more than a day to get free of the muddy flats. This was near where the town of Newstead is today – and there is a story that settlers found the imprint of wheel tracks more than a decade later.

Travelling along the valley near Chewton, Mitchell saw the heights of Mt Alexander. Passing through a ravine they reached a spot near Harcourt where a wheel of the ox-wagon broke, so they made camp. Mitchell took the opportunity to ascend Mt Alexander and from there saw a high range to the south. Next morning he took a small party on horseback with a Koori guide, while the others tended the broken wheel. He hoped to make a sighting of Port Phillip Bay.

They crossed the Coliban near Taradale and after reaching level ground, the party once more saw the mountain to the south. Soils were rich and heavy on the plain, and a second good stream Mitchell named Campaspe. He followed the stream and crossed it once more at the point where Mollison St, Kyneton, crosses it today. At Carlsruhe another crossing of the Campaspe was made and the party headed for the wooded slopes of the mountain he was to name Mt Macedon.

Above: Geoff Hocking, Mt Alexander
Opposite: Pete Walsh, Lake Daylesford

MITCHELL'S EXPEDITION REACHES MACEDON

Major Mitchell approached Mt Macedon from the north, having left the main part of his expedition at Harcourt with a damaged ox-wagon. He found the mountain covered with daisy-bush and wombat holes. The slopes were so thickly timbered, especially on the southern side, that it took a long time for Mitchell to get a viewing point. He only achieved this late in the afternoon, as the light was beginning to go.

With his binoculars he discerned the heads of Port Phillip, and white shapes that he took to be either tents or the sails of a boat. It turns out that on 30 September 1836, a ship from Sydney did sail across the bay, bringing the new administrative party for the settlement at Melbourne. This enterprise had entirely been begun and authorised during Mitchell's absence from Sydney and it was sheer chance he was looking out on the Bay when the official party approached the new village on the Yarra.

Mitchell named the mountain he had climbed Mount Macedon for Philip of Macedon, the great warrior-emperor – apparently relating it to the name of the bay. The next morning, the day being wet and misty, Mitchell's party made for Mt Alexander. Arriving, they found the wagon was repaired and they resumed their journey homeward.

Many towns sprang up along the route of Mitchell's exploration – some by chance, others because Mitchell crossed streams and ranges at points where others later found it convenient to settle. There are monuments in many of these towns, and indeed throughout western and northern Victoria, illustrating how thoroughly Mitchell covered the territory during the course of nine months. In the Heart of Victoria, Mitchell's exploration is remembered at Newstead, Castlemaine, Chewton, Harcourt, Malmsbury, Kyneton, Woodend, Mount Macedon and Bendigo.

After his expedition, Mitchell made a journey back to England with the hope of having his diary records published, and publicising the economic promise of the area now comprising Victoria. He was knighted by a young Queen Victoria, who had begun her reign while he was busy exploring a large part of the colony that later bore her name.

CASTLEMAINE ART GALLERY

The Castlemaine Art Gallery & Historical Museum is one of the most elegant and well-lit regional galleries in Australia. The Gallery has always specialized in Australian art, with separate displays of oils, watercolours, prints and drawings covering most periods in Australian art history. Works by well-known artists from the traditional landscape painting period are a feature of the permanent collection.
14 Lyttleton St. Castlemaine.
Open 10am til 5pm Mon – Fri 12am til 5pm Sat & Sun   **Tel:** (03) 5472 2292
Email: info@castlemainegallery.com   www.castlemainegallery.com

Opposite: Joe Mortelliti, Hanging Rock

BOURKE & WILLS EXPEDITION

The 1860 Burke & Wills expedition from Melbourne to the northern shore of the Australian continent has long passed into legend. Like so many of Australia's heroes, Robert O'Hara Burke and William John Wills cemented their place in our pantheon of heroes by perpetrating a noble failure: they died in a gallant, though tragically inept, attempt to find a route to the north coast of Australia.

The advent of a telegraph link across Europe from London to India in 1859 increased the urgency of finding a route across the Australian continent. Melbourne was linked to both Adelaide and Sydney by telegraph, and if a landline could be established from Australia's northern coast the time for communicating with Europe would be reduced from months to days. The Royal Society of Victoria set up a committee to pursue the goal. With both business and government support, a team of camels was purchased from India and leaders and members were sought for an expedition party.

Robert O'Hara Burke, the Police Superintendent at Castlemaine, was a man who craved excitement. He had successfully quelled a racist riot against Chinese miners at Beechworth, but the life of a country policeman held little promise of action. His unrequited love for the popular actress and singer Julia Matthews made him even more restless. Though not a bushman himself, when he heard of the intended expedition he was determined to be its leader. His glorious return wearing an explorer's laurel would win him fame and the diva's devotion!

Burke was made head of the expedition, and he appointed a serious young astronomer, William Wills, as surveyor. The public took great interest in the project, and over 10,000 people crowded into Royal Park on August 20, 1860 to cheer the expedition on its way. The expedition camped at Essendon the first night, then made its way to Bendigo via Lancefield and Heathcote. The plan was to cross the Murray River near Swan Hill and the Darling near Wentworth, then proceed north to Cooper's Creek; from there to follow the 141st meridian until the northern coast was sighted.

As the expedition traveled through Victoria, thousands of people came to see them pass. There are several claims the team camped camp at villages, got provisions, but then moved off next morning without paying! This allegation was made after their stop beside Deep Creek at Lancefield, and the next day after their overnight stay near Baynton.

Later on, concerned at the pace of the expedition, Burke decided to go forward with a small party consisting of himself, Wills, King and Gray, after arranging a rendezvous with a group led by Brahe at Cooper's Creek. The privations and heroism of the small advance party are legendary. They reached the Gulf of Carpentaria on February 11, 1861, but for the return journey they had only five weeks' supplies. Gray died en route, but the remaining three arrived back at the Cooper's Creek camp on April 21.

Opposite: Alison Pouliot, Mt Franklin

24.

BURKE & WILLS LIONISED

Tragically, the camp at Cooper's Creek was empty when Burke, Wills and King returned – Brahe's backup party had left the previous day, and was fourteen miles away. Following the instruction to DIG (carved famously upon the tree), they found some supplies, and an explanation that Brahe had decided to return to Menindie because his own backup, led by Wright, had failed to arrive.

While Burke, Wills and King set off south-west towards Adelaide, Brahe finally met up with Wright. Brahe and Wright then decided to return to Cooper's Creek hoping to meet the explorers. Although Burke and Wills had buried a message with their intended actions at an agreed spot, Brahe and Wright did not dig to find the message; finding the camp empty they simply turned around and made their way back to Menindie.

Abandoning hope of trekking on to Adelaide, the desperate trio of Burke, Wills and King attempted to return to Cooper's Creek. Wills, forging ahead, found no note or extra provisions left by Brahe or Wright. Local Aboriginals did all they could to care for the wretched men, but only King survived in their care, to be found by a search party in mid-September.

The rescue party, with King, made its way south, cheered by crowds in every township on the way. The diggers of Bendigo fired pistols in the air and showered King with flowers and food. At Castlemaine, where Burke was regarded as a local hero, King was put on a special train for Melbourne.

The dead explorers' remains were accorded vast public respect, and the funeral procession to Melbourne Cemetery was watched by a weeping crowd of 40,000 people. A vast 34-ton monument was built over the bodies, and a public subscription raised a further £4000 to have a statue erected. This stands opposite the Melbourne Town Hall at the intersection of Collins and Swanston Sts.

Other monuments were erected in towns across Victoria. Bendigo has a large memorial in the cemetery, but no monument is grander than the one at Castlemaine where Burke, with his local connection, was regarded with great affection. The Castlemaine memorial is a remarkably tall obelisk on the hill above Mostyn St, built of everlasting grey granite.

Despite its failure, no expedition has more completely captured the imagination of the general public, and search parties sent out in their wake added enormously to European knowledge of the interior of the continent. Burke's chief competitor John McDouall Stuart reached the vicinity of Darwin early in 1862, and as a result the Overland Telegraph Line was built from Adelaide in the 1870s, connecting this most isolated continent to the rest of the world. Another curious outcome was the expedition's demonstration of the ability of camels to survive in the Australian interior. Today there are great herds roaming the wild, descendants of those brought in over the next decade to work in the centre.

Opposite top: Gary Chapman, Harcourt Station
Opposite bottom: Gary Chapman, Harcourt

26.

GOLD!

Towns like Kyneton were founded before the gold rush, but the Heart of Victoria was then essentially a quiet pastoral region. Everything changed forever when gold was discovered in 1851. One hundred and fifty years after the beginning of the gold rush, it is not always easy to comprehend the scale of the diggings, the enormous industrial operations or the vast human numbers attracted by the lure of fortune.

Many thousands came in the first decade, digging and panning the surface, or alluvial, gold. Only a few won any great success - through luck, sheer hard work, and more luck. Then came the building of shafts that accessed reefs in creek banks and hillsides and operations grew from a few individuals working together, to large combines with the capital to acquire and set up mining machinery. By the third decade, in the 1870s, gold mining had chiefly become the preserve of the big operators.

ALLUVIAL MINING

Alluvial mining created the first great gold rush in the 1850s and 1860s. This was the phase when some people literally picked up nuggets off the ground. Miners swarmed over the land with picks and shovels, panning and cradling in the creek bottoms where the action of erosion over the ages had washed

particles of gold, sand and silt into sump deposits. Each miner worked a small claim and dealt with fortune or misfortune as it came to him. More often than not, miners would band together to share the heavy work of developing neighbouring claims. Some got rich, but many succumbed to disease and dysentery. Others faced by reality returned to live by their old skills in the towns and cities.

The best gold seams were found where ancient rivers between 8 and 10 million years ago had washed into the beds of even older streams carrying the gold from a 400 million year old mountain range into fairly concentrated deposits. Miners dug shafts into the layers of ancient riverbeds, carrying the earth solids out to the running water of a nearby creek to wet the material into a slurry so that the heavier flakes of gold would sink to the bottom.

Opposite top: George Rowe, The End of the Rainbow, Golden Square 1857, watercolour on paper, Newson Bequest Fund 2004, Collection: Bendigo Art Gallery
Opposite bottom: Thomas Wright, Sandhurst in 1862, oil on canvas, Gift of the Bendigo City Council 1908, Collection: Bendigo Art Gallery

DEEP LEAD MINING

The practice of deep lead mining - a set of techniques for pursuing the gold seams to great depth underground - was to a great extent developed in this part of the Victorian goldfields. Deep lead mining was a major industrial undertaking, and a secure financial basis and engineering expertise were required to begin operations. In some cases leads were traced beneath hundreds of metres of flow basalt. The challenge was to get to the gold and make a profit on the operation.

The 'deep leads' were quartz seams holding quantities of gold that could be separated when the quartz was crushed. So the seam was followed underground at different levels and the hard rock brought to the surface, where it was baked in kilns to make the quartz brittle. It was then crushed and the gold retrieved.

Bendigo was the first place in the world where mine-shafts reached a depth of one mile underground. The vital technology that enabled these developments was later to be used in the richest goldfield of all – Johannesburg, South Africa.

During the "puddling" phase of the gold rush (the first ten years) at Bendigo, attention was becoming focused on the quartz reefs running under the field. Over time as the reefs were explored in the areas of general mining activity, a picture was built up of their structure. There are thirty-seven lines of reefs running mostly parallel to one another.

These reefs occurred as the result of pressures, which caused folding of the very ancient beds of sand and mud already turned into sandstones and shales over time. During this process, gaps and cracks opened between the folding layers of rock. These openings became filled with solutions under immense pressure, solutions that carried with them gold and pyrites (fool's gold). When these had solidified through slow cooling, they formed the crystalline rock known as quartz along the crevices. The ancient beds had been tipped over from their original horizontal position. So the quartz reefs have a pattern that follows the line of folding and tilting of the stone beds.

The processes that brought the gold into the folded and tilted beds of stone did not then cease, but continued to cause further changes, with later intrusions of fluid rock material. The task of the goldminer, aided by the mineral geologist, was to discern in the rock structure the pattern of the reefs, by following the complex evidences of the rock-materials encountered through a vertical bore into the earth. Then a shaft was sunk, from which tunnels were driven laterally to locate the reef.

The expense of such mining processes lay first in setting up machinery to provide access to the shaft through raising and lowering the elevator cage. This would bring the workforce to the job, and bring the waste rock and mud to the surface for disposal. At any depth, water was likely to enter the tunnels, so pumping equipment was essential. This was driven by a steam engine (wood-fired), which provided motive power for the elevator also. Outside the mine, mullock heaps had to be established clear of the workings, then kilns to fire the quartz to remove associated minerals. A crushing plant would then break up the quartz. The residue was washed and shaken so that gold would separate out. Conditions for the miners and others workers on site were very tough and uncomfortable, with huge amounts of dust and mud in the tunnels, and smoke and filth in the air at the surface workings. An experience of this kind of mining (without the filth and dust) can be gained at the Central Deborah Mine near the city-centre of Bendigo.

Opposite top: Katherine Seppings, Chewton
Bottom: Katherine Seppings, Castlemaine Diggings

DEMOCRACY

The greatest social and political movements of the 19th century were stimulated by notions of freedom and equal opportunity for all. These notions arose largely from the great revolutionary movements of the previous century, and were highlighted in social debate at the time of the gold rush. Most of those who arrived to seek their fortune with pick, shovel and pan were men of some means, at least able to afford their passage to Australia. Certain principles and ideals were in the air:

- the right of each man to have a vote in the government of the state,
- the right of education for all,
- freedom of expression as exemplified by a free press and unhindered public assembly.

With such commonly shared notions on the goldfields, there was keen debate on the government's levy on mining rights and any proposed taxes. Agitation Hill in Castlemaine is a famous site where diggers grouped together to air grievances. Local historians see these gatherings, together with those on other goldfields such as the Monster Meeting at Chewton and Red Ribbon Day in Bendigo, as the beginnings of Social Democracy in Australia, leading to the ultimate expression of the Eureka Stockade in Ballarat three years later.

Agitation Hill was just across Barkers Creek from the Government Camp Reserve, which housed the Goldfields Commissioner's Office, the Police Barracks and the Courthouse. Miners gathered on Agitation Hill, where they had a visual advantage over the Government Camp Reserve, to express their complaints. The miner's complaints were chiefly over the manner of issuing mining and liquor licences, and the habitual cronyism among officers of the law and some local businessmen. Many of the agitators were dismayed that the Church of England was allotted the site when the town was surveyed in 1853.

Every town had a Mechanics Institute - part of a social movement finding a foothold in every new community in the country during the mid-19th century. Mechanic's Institutes and Schools of Mines aimed to offer the working-man facilities for self-improvement. The principles of trade unionism and universal franchise were also at the heart of the goldfields. Temperance Halls also appeared, although they were quite somewhat outnumbered by hotels – there were 400 hotels on the Bendigo goldfields.

Mechanics Institute halls and reading-rooms were erected by public subscription. Some became the site for future local government – the shire or town hall – or municipal libraries; others have grown to form a part of the nation's tertiary education system.

Above top: Katherine Seppings, Castlemaine
Above bottom: Katherine Seppings, Mechanic's Institute, Malmsbury
Opposite top: Gary Chapman, Castlemaine Botanic Gardens
Opposite bottom: Gary Chapman, Castlemaine Botanic Gardens

ENGINEERING

RAILWAYS

The Melbourne-Bendigo railway, built between 1860 and 1862, was the largest public work in the colony in the 19th century, with a set of bridges and stations consistently designed with style and purpose.

It was the first double track on the Victorian country system. Stations and ancillary sheds, bridges and viaducts are all solidly built in brick and stone. There are two tunnels – a scarce feature on the Victorian system. Most of the bridges are masonry-arched and some road-bridges over the line are of riveted steel beams – the latest thing in the 1860s. Two in particular are worthy of mention as great railway structures, the Malmsbury Viaduct and the Taradale Bridge.

Malmsbury Viaduct is a massive structure of bluestone arches built across the deep valley of the Coliban River, just before Malmsbury Station. It is easily seen from the Botanic Gardens and the Calder, and can be approached by turning off the highway at Malmsbury Town Hall.

The magnificent bridge at Taradale consists of alternate stone and steel piers supporting a steel-trestle deck. The bridge is forty metres above Back Creek, and was for many years the highest on the Victorian railway system. The road that leads beside the creek should be followed as far as the base of these great piers, so that its true scale can be appreciated.

WATER TO THE GOLDFIELDS

The main element required for the recovery of alluvial gold is water – large amounts are needed to separate the heavier flakes of gold from mud and gravel.

Because of the reliable flow of the Coliban and Little Coliban rivers, a project was developed to channel water to the valleys around Castlemaine and to Bendigo. The water was to be used for sluicing the hillsides and old mine workings for residual alluvial gold.

The Victorian Government took on the expense of this huge scheme in the late 1860s. A large dam was built on the Coliban River above Malmsbury to bring water by a system of gravity-fed channels to the gullies of Mount Alexander Diggings. The main channel continued for a distance of 70km to the Bendigo Goldfield where the Sandhurst Reservoir was opened in 1874, filling up by 1877.

The Malmsbury scheme was the largest water conservation system in Australia in the 19th century. The dam wall is directly up-river from the railway viaduct and can be seen from trains approaching the station. The reservoir can be reached by road, taking the Daylesford road from the Calder Highway in Malmsbury, turning left after the railway bridge and proceeding to the gate into the picnic ground.

Malmsbury Dam has a long earthen embankment faced with basalt blocks, with concrete spillways at either end. From there a channel runs beside the river, underneath the great stone railway viaduct. By the time it flows under the Calder, the channel moving on its contour is already quite a distance from the river, by then descending into a deepening valley. Towards Elphinstone the water-channel passes under the road, and can be seen winding around the hillsides on its contour. Subsidiary channels are encountered in the old workings around Chewton, Fryerstown and Castlemaine.

In the 1960s Lake Eppalock was established 20 km north where the Coliban joins the Campaspe River. Both rivers have by then descended about 500 metres from their source on the Dividing Range near Trentham, and from the high tableland country near Kyneton and Malmsbury. The lake itself was dammed to provide water for the growing towns of the Bendigo region and for irrigation on the plains to the north.

Opposite top: Daryl Chibnall, Malmsbury Viaduct
Opposite bottom: Daryl Chibnall, Malmsbury Dam

GARDENS

Many Victorian botanic gardens were given plants and suggestions by Baron Ferdinand von Mueller, the inaugural director of the Melbourne Botanic Gardens. One of the great botanists of the 19th century, he corresponded with other botanists throughout the world. He travelled energetically and listed and classified vast numbers of Australian plants. He was not, however, a landscape designer, preferring to plant related species in rows like the true scientist that he was. There is little doubt however, that the baron, in his day, inspired many to take plant-research and gardening seriously. This was a spur to the development of botanic gardens throughout Victoria, even in quite small country towns.

The **Kyneton Botanic Gardens**, near the historic 1870s bridge, are among the finest in the state, with over 1000 trees along the northern bank of the river. Planting began in 1863, helped on by a gift of flowers and shrubs from Baron von Mueller.

Laid out on the east bank of the Coliban River, the **Malmsbury Botanic Gardens** were opened in 1859. The fine collection of large conifers and deciduous trees is interspersed with mature shrubs, especially arbutus. The lake has a prominent bluestone fountain and breeding populations of waterbirds. There is a magnificent view up river to the great bluestone arches of the 1861 railway viaduct. Malmsbury may well be the smallest town in the world to have its own botanic gardens.

The beautiful **Daylesford Botanic Gardens** occupy the crest of Wombat Hill, an extinct volcano, giving energetic growth to a wide variety of conifers and flowering shrubs. Firs and redwoods from Europe and North America are the chief glories of the gardens. The nine hectare gardens were originally established in 1861 and, with various additions made over the years, they are among the best in the state.

The **Castlemaine Botanical Gardens** is situated away from the centre of town, in Walker St. Monumental pillars, a carriage drive and large lake allow the visitor to experience a 19th century designed landscape. The reserve was set aside in 1860. One of the oldest known cultivated trees in the state - an English Oak planted in 1867 - still thrives here.

Also in Castlemaine, **Buda Historic Home & Garden** in Hunter St is a cool retreat on a hot summer's day. The grounds surrounding Buda date back to the 19th century, with most of the plants brought into the garden in the 1860's remaining. This collection includes elaborate Victorian-style flower beds, formal gardens reflecting early 20th century garden design, a garden pavilion and an ornate aviary.

In Bendigo, **Rosalind Park** occupies one of the city's best vantage points and is a historical and natural landmark near the city-centre. After strolling through an open air fernery and elegant conservatory, and seeing the wonderful 'rustic' cascade, the lookout tower offers the visitor a spectacular view of the town. The **Rose Lee Gardens** at Kangaroo Flat is one of the finest rose gardens in the Goldfields, These relaxing, fragrant and colourful gardens are at their peak from December through to April. The Chinese contribution to the history of the region can be seen in the **Classic Gardens & Kuan Yin Temple**. In the Chinese culture, water gardens seek to create a balance between rocks and water, according to the philosophy of yin and yang (male and female).

Top: Daryl Chibnall, Buda
Opposite: Alison Pouliot, Daylesford Botanic Gardens

EVENTS & FESTIVALS

Throughout the Heart of Victoria local events and festivals celebrate the region's history, food and wine, or showcase the arts. New ideas and energetic local initiatives mean new concerts or exhibitions are happening all the time. For up to the minute ideas of where to go and what to see, contact the local visitor information centres (see separate section).

The **Lavender Harvest Picnic** in Shepherds Flat, near Hepburn Springs in January is a great event that showcases local produce and celebrates the lavender harvest.

The famous **Harvest Picnic at Hanging Rock** is held each February, and you can join the **World's Longest Lunch** celebrations starting at Hanging Rock in March. At easter see **Sun Loong,** the Chinese dragon, parade through the streets of Bendigo for the annual six-day festival.

Near Hepburn Springs, the **Swiss Italian Festa** – a celebration of the unique Swiss-Italian heritage of the Hepburn Springs and Daylesford area – offers a myriad of food and wine events, musical and theatrical performances and cultural festivities in May.

In October tour Bendigo's heritage buildings and taste entrees from local restaurants accompanied by the best of the region's wines, at **Bendigo Heritage Uncorked**. The **Budburst Festival** in the Macedon Ranges wine region is a chance to experience the best of the region's wines and food and the leisurely local pace.

The **Maldon Folk Festival**, a four-day musical extravaganza held over the Melbourne Cup weekend in November, features Australia's finest folk artists.

Horse races are popular events in this region, with many regular meetings. The picnic meetings are famous, particularly the **Hanging Rock Races** and the **Kyneton Cup**.

Above: Geoff Hocking, Maldon
Opposite top: Katherine Seppings, Cars, Market building Castlemaine
Opposite bottom: Alison Pouliot, Lavender

EVENTS & FESTIVALS

JANUARY
Boite Singers Festival – Daylesford

Hanging Rock Races

Lavender Harvest Picnic – Shepherds Flat

Bendigo Country Week Cricket

FEBRUARY
Harvest Picnic – Hanging Rock

Daylesford Macedon Produce Day

Film screening of Picnic at Hanging Rock – Hanging Rock

MARCH
Bendigo International Madison – Bendigo

Harcourt Applefest – Harcourt

Gear Grinders Motorfest – Maldon

Begonia Display – Daylesford

Chill Out – Daylesford

Hanging Rock Races

World's Longest Lunch – Hanging Rock

Castlemaine State Festival

Bendigo Easter Festival

Bendigo Wine Festival

Bendigo Antique Fair

Chrysanthemum Championships

APRIL
Anderson's Mill: Food, Wine and All That Jazz – Smeaton

Mt Macedon Autumn Display & Open Gardens

Bendigo Easter Festival – Bendigo

Castlemaine State Festival – Castlemaine

Maldon Easter Fair – Maldon

MAY
Mt Macedon Open Gardens Scheme

Swiss Italian Festa – Shepherds Flat

JUNE
Bendigo Eisteddfod

JULY
Glenlyon Fine Food and Wine Fayre

AUGUST
Words in Winter Festival – Daylesford and Hepburn Springs

SEPTEMBER
Kyneton Daffodil Festival

Mt Macedon Open Gardens Scheme

OCTOBER
Bendigo Heritage Uncorked – Bendigo

Mt Tarrengower Hill Climb – Maldon

Macedon Ranges Budburst Celebration

Mt Macedon Open Gardens Scheme

NOVEMBER
Daylesford Agricultural Show

Fabulous Flower Fiesta – Hepburn Springs

Macedon Ranges Open Gardens

Castlemaine and District Festival of Gardens – Castlemaine

Maldon Folk Festival – Maldon

The Kyneton Cup

Mt Macedon Open Gardens Scheme

DECEMBER
Daylesford Highland Gathering

Daylesford New Year's Eve Gala

Opposite: Alison Pouliot, Glenlyon

WALKING

The area offers a great variety of walks from strolls through gorgeous gardens to strenuous hikes, including the outstanding Great Dividing Trail.

Among the most popular walking tracks are the **Tipperary Walking Track**, Daylesford, a 16 km walk that takes you past the remains of water races used by gold miners in the last century, through native bush and stands of European trees and past mineral springs. From the outlet of Lake Daylesford it leads through Central Springs Reserve and along Wombat Creek.

For a view, the **Hanging Rock Trail** is unbeatable. Hanging Rock was made famous as the subject of the classic Australian novel and movie 'Picnic at Hanging Rock'. Climb through the rocky maze to the 718-metre summit for a bird's-eye view over the surrounding plains and farmland, and nearby Macedon and Cobaw ranges.

To learn about the environment, the **Forest Ecotourism Walking Trail** at Mount Macedon is a great family walk that seeks to educate visitors about the native forests of the Macedon Ranges. See diverse plants that provide habitats for over 150 species of native birds and fauna, and loop past the peaceful tree-fringed Sanatorium Lake.

Good bushwalking can be found in all the national, state and regional parks in the area, including the **Lerderderg State Park**, **Macedon Regional Park**, **Hepburn Regional Park** and **Hanging Rock Reserve** as well as the **Wombat State Forest**, with its fern gullies, streams, waterfalls and spring wildflowers.

Above top: Sandy Scheltema, Lake Daylesford
Above bottom: Joe Mortelliti, Wombat Forest
Opposite: Joe Mortelliti, Rosebury Hill, Pipers Creek

THE GREAT DIVIDING TRAIL

The Great Dividing Trail follows the Great Dividing Ranges and reveals Central Victoria's mining relics and its natural beauty to walkers – from deep gorges and fern-lined rivers to the artifacts and cultural heritage of the greatest gold rush the world has ever seen.

The trail is a 260 km public walking trail following the top of the Great Dividing Range from Bacchus Marsh to Bendigo. Created by a community-owned organisation, the Great Dividing Trail Association, the trail links the old gold rush towns at the heart of Victoria, as well as the forests, hills and lakes, straddling the Great Dividing Range.

The Great Dividing Trail consists of four interconnecting major tracks: **Dry Diggings**, **Federation**, **Lerderderg** and **Leanganook** tracks. Each major track consists of minor thematic walks (such as the **Miners Walk**), which can be easily accessed for day walks. The Great Dividing Trail Association is currently completing the final extension of the walking track to Bacchus Marsh from Daylesford via the Lerderderg Track. Track completion in 2004 will link Bendigo and Ballarat with a walking track to the edge of Melbourne for the first time since the days of the gold rush of 1851.

The first track that was completed was the **Dry Diggings Track** from Castlemaine to Daylesford. This is a 55-km walking route which winds its way around the old goldfields between Castlemaine and Daylesford, taking in Fryerstown, Vaughan, Mt Franklin and Hepburn Springs. It takes in many of the area's gold mining relics, as well as its plant communities and fauna types.

For more information see the association's website at www.gdt.org.au.

Above: Katherine Seppings, Castlemaine Diggings, Welsh Village
Opposite: Katherine Seppings, Chewton

PICNIC SPOTS

The parks and gardens of the region offer wonderful places to picnic, and there are also plenty of spots that are bit less cultivated, where a kangaroo or wallaby just might hop by.

Lake Daylesford is a popular fishing spot with picnic-barbecue facilities. The peace mile walking track circles the lake. Rowboats, aquabikes, paddleboats, canoes and surf skis can be hired on the foreshore. **Jubilee Lake**, at the south-eastern edge of Daylesford was constructed in 1860 to supply water for the goldfields and for domestic purposes. It is now a popular spot for boating, picnicking and swimming. There is a caravan park, a mineral spring, a kiosk, barbecue facilities, boat and canoe hire. A walking track leads around the lake and on to Soda Spring.

Sailor's Falls are situated in an area that was once mined for gold, about 8 km south on the Daylesford-Ballan Rd. They cascade for 30 metres down a steep gorge into a fern-lined creek. There are picnic facilities and electric barbecues. A short loop walk takes in the area's mineral springs or you can follow the orange trail markers and signposts 6.5 km to Twin Bridges.

Mt Franklin Recreation Reserve on an extinct volcano, is an unmistakable landmark 10 km north of Daylesford along the Midland Highway. The road ascends to a sizeable and shady picnic-barbecue area. Short-term camping is permitted with fireplaces, toilets and washing water provided. You can walk or drive from the picnic area to the excellent lookout at the summit.

Lake Eppalock on the Campaspe River, 26 km east of the City of Greater Bendigo, was constructed between 1960 and 1964 to provide irrigation supplies from the Campaspe River, and the town water supply to Bendigo. Boating (power and sail), water skiing, swimming and fishing take place, and facilities around the lake include caravan parks, camping grounds, picnic areas and public boat ramps.

Above: Pete Walsh, Daylesford
Opposite: Pete Walsh, Lake Daylesford

ACTIVITIES

HERITAGE TRANSPORT

A number of the historic towns in the region have various forms of heritage transport. In Daylesford train buffs take a vintage diesel rail motor line through the Wombat State Forest, and on the first Saturday of each month board the Silver Streak Champagne run from Daylesford Railway Station at 5.30 pm. Drinks and finger food are served on the train.

The **Victorian Goldfields Railway** in Maldon (look for the station in Hornsby St) allows the traveller to enjoy the sights sounds and atmosphere of the time when steam ruled the rails.

Interstate and overseas visitors to Bendigo particularly enjoy the **The Vintage Talking Tram** tour from the Central Deborah Mine along Bendigo's main street. At the Central Deborah Mine itself, you can don a helmet and a miner's lamp and descend underground for more than 60 metres (over 30 storeys) to see how gold was - and still is - extracted from the quartz reefs running deep beneath Bendigo.

HORSERIDING

Horse riding is available at Tallara, on Spring Hill Rd, Daylesford. There's Hill's Pony Playgroup, and Lead Trail Rides for those aged two to twelve (for groups of 2 or more) and private lessons for beginners to advanced. Burrinjuck Trail Rides at Burrinjuck Park, Spring Hill, is open on weekends and public holidays for two-day trail rides in the Wombat State Forest.

SHOPPING

Prospectors still pick over the hills of the Heart of Victoria, but treasure hunters also search the region's many antique and junk shops. There's furniture, bric-a-brac, silver, china, and hardware. Antique shops worth exploring include the three storey **Macedonia** antique centre in Lancefield. Kyneton's Piper St is also renowned for its antiques and has other highlights such as the **Kyneton Fine China** shop, for handcrafted ceramics. There are an increasing number of contemporary craftspeople making furniture, sculptures and other decorative features.

There are also many fine second-hand bookshops – in fact there is a leaflet describing a trail of second hand bookshops through the Heart of Victoria available from visitor information centres.

This is spa country – the home of mineral waters, mineral baths and masseurs – so perfumed salts, pure vegetable oil soaps, bombs and bubbles, floating candles, scented candles, flower essences, natural cosmetics, and hair and skin treatments are in specialty shops everywhere.

The region has a wonderful collection of clothing shops, including **Meskills Woolstore** selling pure wool, knitting yarn and knitwear in Kyneton. Nearby Woodend also has a number interesting shops, including **Cottage Courtyards** and the **Literary Latte**, which is a great place to browse while sipping on a coffee.

Above: Pete Walsh, Daylesford
Opposite: Sandy Scheltema

NURSERIES

Plant nurseries abound through the Heart of Victoria and in summer locals and visitors stroll and marvel at **Romswood Peony Farm**, Kerrie. Steven Ryan's **Dicksonia** at Mount Macedon has an astonishing repertoire of exotic species, while Kathy at **Florelegia** has hellebores each year to make the loveliest garden. **Macedon Nursery** offers a broad range of plants, especially cool climate species for the local enthusiast.

MINERAL SPRINGS

The 'Daylesford and District Mineral Springs Map' is available free from the Daylesford Regional Visitor Information Centre. Visitors are encouraged to sample the therapeutic waters of the springs via old-fashioned hand pumps, or continuous flowing pipes, and to bring receptacles to be filled for home use at the pumps in various springs.

SPAS

Numerous spas in the area offer every known method for rejuvenating tired bodies. Because of its reputation for health and well-being the area has probably the greatest concentration of masseurs and natural therapists in Australia.

Private spas offer aromatherapy and homeopathic or Australian bush flower essences are added to baths to scent the water and increase their therapeutic benefits. Spa couches with aero jets give a powerful water massage, flotation tanks cocoon from the pressures of the outside world, and heavy mineral salt pools allow the body to relax as it feels weightless. It is possible to arrange for a masseur to come to your cottage or country retreat.

The Springs Retreat, Main Road, Hepburn Springs offers products that are based on ancient Chinese methods for refreshing the mind, body and spirit. **Aqua Viva Day Spa** has luxurious packages including massages, facial massages, hydrotherapy, mineral baths, body scrubs, golden moor mud wraps, vichy showers, and steam baths. **Shizuka Ryokan Day Spa Retreat** Hepburn Springs is a traditional Japanese environment, offering reiki, shiatsu or relaxation massage. The **Hepburn Spa Resort** is Australia's main mineral water spa and offers a full range of treatments from spas baths, massages, mud wraps, facials, and flotation tanks. There are many more specialised spas offering luxury and relaxation for all tastes.

Above: Pete Walsh, Sunday Market Daylesford
Opposite: Greg 'Arjuna' Govinda, Lake Daylesford

PLACES TO STAY

The area has a broad range of accommodation for all budgets and tastes. Choose from historic and boutique B&Bs, 19th century miners' cottages or grand old hotels with ornate Victorian architecture. The **Shamrock Hotel** in Bendigo offers the visitor a taste of bygone days. You can choose from two restored bishop's palaces: **Langley House B&B** or old **Bishopscourt**.

The elegant, modern **Blue Ridge Inn** on Mount Macedon has a great view, looking across the plains with Hanging Rock mysterious in the middle ground. **Campaspe House** and **The Bentinck** in Woodend offer 1st class accommodation in a restored arts and craft-style style mansion. **Villa Parma Pensione**, in Hepburn Springs, is a historic Italianate house surrounded by an Italian-style garden.

Families are well catered for too with a range of holiday parks and campsites throughout the region. Or you can experience life on a working farm at one of the many farm stays.

See the **Business Listings** section at the back of this book for an extensive directory of accommodation.

CONVENT GALLERY

A place of Historic, Aesthetic & Spiritual Seduction the impressive CONVENT GALLERY is renowned for its suberb art, food and atmosphere. An art lover's haven, the award winning gallery also houses a Mediterranean-style restaurant, a stylish bar, restored chapel, museum and an extensive retail level displaying jewellery, designer clothing, local produce and unique gifts. A feast for all the senses.

Daly Street. Daylesford 3460. **Open** 10am til 5pm Daily
Tel: (03) 5348 3211   **Fax:** (03) 5348 3339
Email: visit@theconvent.com.au   www.theconvent.com.au

Opposite: Pete Walsh, Daylesford

FOOD

EATERIES

This region has a reputation for culinary excellence, with a number of local eating establishments featured in The Age Good Food Guide. You can often sample the local produce at these excellent local restaurants.

Notable examples include the **Lake House**, Daylesford, recognised as the 'Country Restaurant of the Year' in the Age Good Food Guide 2004 and **La Trattoria** at the Lavandula Lavender Farm near Hepburn. **Bazzani** in Bendigo, **The Globe** in Castlemaine, and **Café Colenso** in Woodend are recommended by discriminating locals who enjoy dining out.

As the area develops, more and more chefs are opting for a country lifestyle, so the range of good dining experiences continues to expand. For cosy dining by log fires in winter visit country retreats such as the **Macedon Family Hotel**, an unpretentious building which surprises with food and service, **Holgate's** at Woodend, or the **Thai Issan** in Trentham.

Cafés in towns and villages offer city-quality espresso and meals served in a laid-back country atmosphere. Try **Frangos** in Daylesford, **Café Colenso** in Woodend, **Gonnella's** in Kyneton, **Barzurk** in Bendigo, **The Chimney** in Trentham, and **Saff's** in Castlemaine.

See the **Business Listings** section at the back of this book for an extensive directory of eateries.

LOCAL PRODUCE

An increasing range of superb local products are now on offer in the region. Many farmers are now growing organic produce ranging from herbs and vegetables to grains and fruit.

In summer look for succulent raspberries and blueberries, such as those from the **Woodend Berry Farm** on the Tylden road, and for blackberries, currants, and cherries, stone fruit and yabbies.

In autumn you'll find apples, sweet chestnuts and wild mushrooms. In spring you might spot the speciality morel mushrooms and the first new season baby vegetables, while sacks full of freshly dug potatoes such as pontiac, Idaho, kipfler, kennebee and desiree appear throughout the year.

For olives, and olive oil infused with lemon or herbs, and speciality tapenade, try Kyneton's **Barfold** olive oil, on the Heathcote Road. You can buy fresh hazelnuts on the Barringo Road to Mount Macedon.

Above top: Pete Walsh, Hepburn Springs
Above bottom: Pete Walsh, Main St Glenlyon
Opposite: Sandy Scheltema, Trentham

Stop at roadside stalls where farmers sell apples, tomatoes, eggs or berries or call in at local butchers and delicatessens to buy goods such as fresh and smoked trout, yabbies and spicy sausages, along with a wide range of preserves, chutneys, pickles, jams, honey, and flavoured vinegars. Look out for local produce signs as you travel around the region, and carry some coins to make the correct change when you find an honesty box.

While most of the larger towns and villages have supermarkets, the village butcher, often specialising in locally grown lamb, venison and other game, is still very much part of the region. They will bone and prepare a cut and chat the while; ask for some home made sausages or find a wild rabbit (underground mutton, as the Woodend organic butcher calls it) to casserole with herbs from the nearby nursery and some local wine.

The annual Harvest Picnic held in February at Hanging Rock, the Farmer's Markets at Lancefield and Kyneton, and the Daylesford Market at the station on Sundays showcase much of the best produce of the region.

Sweet Decadence, Daylesford, specialises in hand made chocolates, and on the Midland Highway near Mount Franklin the **Chocolate Factory** demonstrates the process of making chocolate and offers treats such as chilli chocolate. **Lavandula**, at Shepherd's Flat near Daylesford, sells lavender, olives, herbs, grapes, and arts and crafts from an historic stone building that was built in 1850s, and has a café open daily in summer and at weekends in the colder months.

The **Himalaya Bakery** in Daylesford specialises in breads made according to strict biodynamic principles and using organic sourdough techniques. On the Midland Highway **Country Cuisine's** jams, chutneys, dressings, sauces, and pasta sauces make local catering easy. **Istra Smallgoods** in High St, Musk, has prosciutto, pancetta, capocollo and sausages from locally-bred pigs.

The Nutty Fruit Farm in Glenlyon is worth a seasonal visit for mulberries, hazelnuts, walnuts, apples, pears, chestnuts, organic mesclun and kipfler potatoes. **Tuki Trout Farm** at Stoney Rises, Smeaton is about lamb as well as trout, while **Kyneton Provender** on Piper St will provide Red Kelpie relish, to go with the roast. **Malowa House Gourmet Delights** in Woodend does indeed delight locals with a selection of delicacies from Australia and overseas — there's no need to travel for French butter and cheese, nor even to Carlton for gourmet delights such as fresh pasta and Phillipa's breads.

Above: Pete Walsh, Hepburn Springs
Opposite: Sandy Scheltema, Daylesford

WINE

The **Macedon Ranges Wine Region** is the coolest grape growing climate in mainland Australia and has a reputation for fine sparkling wines. Its altitude of about 500 metres and the resulting cool climate suit chardonnay and pinot noir grape growing, much of which is used in the production of sparkling wine. Many producers market their sparkling under the name 'Macedon', resulting in growing recognition of the region's prominence in the premium sparkling wine market.

Quality table wine remains the backbone of the region, however, particularly shiraz and riesling. Local riesling and cool-climate reds have won major awards and attract a growing band of wine drinkers looking for something different to the jammier, sweeter and more heavily bodied red wines of hotter regions.

The red soils and warm temperatures of **Bendigo Wine Region** produce award-winning shiraz and cabernet wines – all deep red, ripe and boasting a distinct minty edge to the berry fruit. As with other Victorian regions, the early vineyards in Bendigo fell victim to both phylloxera and the economic crash and it was not until the late 1960s that a new vineyard was planted by local identity, Stuart Anderson. Many followed his example, with shiraz and cabernet sauvignon the most common plantings.

BALGOWNIE ESTATE

An award-winning winery established in 1969. The cellar door outlet boasts a wine tasting and sales area, café and wine museum.
Hermitage Road, Maiden Gully
Ph +61 3 5449 6222

BASALT RIDGE WINERY

Offering a number of award-winning wines including superb examples of cool-climate shiraz and merlot, together with chardonnay, cabernet sauvignon and pinot gris.
199 Zig Zag Road, Malmsbury
Ph +61 3 5423 9108

BIG HILL VINEYARD

Set in natural surroundings overlooking shiraz vines and a breathtaking garden setting.
Corner Calder Highway and Belvoir Park Road, Bendigo
Ph +61 03 5435 3366

BIG SHED WINES

Established in 1999 and specialising in pinot noir. The Big Shed is a local landmark on the Daylesford-Malmsbury Road between Glenlyon and Denver.
1289 Malmsbury-Daylesford Road, Glenlyon
Ph +61 3 5348 7825

BLACKJACK VINEYARDS

Specialising in big reds, the cellar door is well worth a visit, but ring before hand as it is only open while stocks are available.
Calder Highway, Harcourt
Ph +61 3 5474 2355

Above: Geoff Hocking, vineyard
Opposite: Sandy Scheltema, Hanging Rock

BLANCHE BARKLY WINES

The big beautiful reds of Blanche Barkly are hand crafted from the low yield Kingower Vineyard for complexity, softness and flavour.
Rheola Road, Kingower
Ph +613 5438 8223

CHATEAU LEAMON

Chateau Leamon produced its first wine in 1977 and quickly established a reputation for premium wines.
5528 Calder Highway, Bendigo
Ph +61 3 5447 7995

CLEVELAND WINERY

Enjoy the historic surrounds, Cleveland homestead was built in 1889, as you taste high quality European-style cold climate table wines.
55 Shannons Road, Lancefield
Ph +61 3 5429 9000

COBAW RIDGE

Everything at Cobaw Ridge is the work of Alan and Nelly Cooper, from the planting of each vine to the building of the mud brick house and winery.
31 Perc Boyers Lane, East Pastoria
Ph +61 3 5423 5227

CONNOR PARK WINERY

Wines are handcrafted using traditional winemaking methods with the least possible chemical intervention from the vineyard to the bottle.
Connor Road, Leichardt
Ph +61 3 5437 5234

COPE-WILLIAMS WINERY

A boutique winery with village green cricket, a royal tennis court and fine dining. The wines are distinctive and a great example of variety and regional character.
Glenfern Road, Romsey
Ph +61 3 5429 5428

ELLENDER ESTATE

Discover Jenny and Graham Ellender's secluded vineyard and winery set in the tranquil and picturesque hills surrounding Glenlyon – a short escape from Daylesford.
260 Green Gully Road, Glenlyon
Ph +61 3 5348 7785

GISBORNE PEAK WINES

Try the award-winning cool-climate wines and enjoy a wood-fired pizza at the weekend all in a most beautiful setting amongst the vines.
69 Short Road, Gisborne south
Ph +61 3 5428 2228

GLEN ERIN VINEYARD RETREAT

Cellar door sales of Gold Medal winning sparkling and other estate grown red and white wine. Boutique Hotel and "Grange" Restaurant.
Rochford Road, Lancefield
Ph +61 3 5429 1041

Above: Geoff Hocking, Vineyard
Opposite: Gary Chapman, Faraday Vineyard
Opposite: Gary Chapman, Faraday Vineyard

HANGING ROCK WINERY

Established in 1983 by John and Ann Ellis, the flagship product is the sparkling 'Macedon', a Champagne style sparkling unashamedly modelled on Krug and Bollinger.

88 Jim Road, Newham

Ph +61 3 5427 0542

GRANITE HILLS

On the north side of the Great Dividing Range, pioneer cool climate winemaker Lew Knight specialises in producing premium cool-climate wine styles.

1481 Burke & Wills Track, Baynton

Ph +61 3 5423 7264

KYNETON RIDGE ESTATE

A family run vineyard established in 1997 and specialising in premium pinot noir. B&B accommodation set in beautiful surrounds.

90 Blackhill School Road, Kyneton

Ph +61 3 5422 7377

LANGANOOK WINES

A small secluded vineyard and winery in Sutton Grange producing premium table wines. The vines were planted in 1985 and are unirrigated.

91 McKittericks Road, Sutton Grange

Ph +61 3 5474 8250

LYNNEVALE ESTATE

A winery and modern homestead style retreat located in the heart of the Bendigo wine growing region.

83 Cahills Road, Bendigo

Ph +61 3 5439 3635

MANDURANG VALLEY WINES

Home grown, hand picked, estate vintaged and bottled high quality wines. We have a passion for making clean wines where concentrated fruit flavors are able to speak clearly.

77 Fadersons Lane, Mandurang

Ph +61 3 5439 5367

MOUNT MACEDON WINERY

A true cool climate vineyard and one of the most elevated in Australia, ideally situated for the production of sparkling wine and fine, cool climate table wines.

433 Bawden (Douglas) Road, Mount Macedon

Ph +61 3 5427 2735

MOUNT WILLIAM WINERY

Continuing to expand in viticulture and product each year, Mount William takes pleasure in producing wines of quality for your enjoyment.

890 Mt William Road, Tantaraboo

Ph +61 3 5429 1595

NUGGETTY VINEYARD

At the base of the Nuggety Ranges on the outskirts of historic Maldon. The vineyard specialises in shiraz and also produces cabernet sauvignon and semillon.

280 Maldon-Shelbourne Road, Maldon

Ph +61 3 5475 1347

ROSEBERY HILL VINEYARD

Rosebery Hill was established in 1984, with the first vintage produced in 1990. Located in a scenic landscape at Pipers Creek, on the north facing slope of an extinct volcano, the vineyard has 20 acres of cabernet sauvignon, cabernet franc and merlot.
Pastoria Rd, Pipers Creek
Ph +61 3 5423 5253

PASSING CLOUDS

A small hands on winery using traditional techniques. Low rainfall and no irrigation combined with well drained deep soil produce wines of unique flavour.
Kurting Road, Kingower
Ph +61 3 5438 8257

PORTREE VINEYARD

Fine Australian cool climate wines from Victoria's Macedon Ranges. Just a half hour's easy drive north of Melbourne Airport.
72 Powells Track, Lancefield
Ph +61 3 5429 1422

SANDHURST RIDGE

A small family operated vineyard nestled against a state forest in rural Marong. Taste award-winning wines in a relaxed cellar door environment.
156 Forest Drive, Marong
Ph +61 3 5435 2534

WATER WHEEL VINEYARDS

Situated on the banks of the Loddon River. Watch winemaking in progress from the cellar door surrounded by hectares of thriving vines.
Raywood Road, Bridgewater on Loddon
Ph +61 3 5437 3060

WELSHMANS REEF VINEYARD

A boutique winery producing full bodied flavoursome wines from estate grown fruit. Tasting and sales are available at the mudbrick cellar door.
Maldon-Newstead Road, Welshmans Reef
Ph +61 3 5476 2733

Above left: Sandy Scheltema, Hanging Rock Vineyard
Above right: Geoff Hocking, Vineyard
Opposite: Pete Walsh, Vineyard Blampied

64.

THE ARTS
BENDIGO ART GALLERY

Bendigo Art Gallery, founded in 1887, is one of Australia's oldest galleries. The collection includes important art from the Bendigo goldfields. Works by S.T. Gill, Thomas Ham, Louis Buvelot, Eugene Von Guerard, Tom Roberts and Arthur Streeton, give strength to the outstanding early Australian collection.

These works are housed in a beautiful series of restored 19th century spaces with polished wood floors, ornate plaster arches and cornices, and diffused natural sky-lighting through rooftop lanterns. There's also a new contemporary wing, including an excellent a restaurant, and a dedicated works on paper gallery. The contemporary building allows flexible spaces for travelling exhibitions.

Elements of the gallery's extensive collection of 19th century European paintings, sculptures and decorative arts are seen by some as sentimental and overly romantic. However, the work is often technically impressive and it can throw throw fascinating light on a generation and culture that had very different reference points. These pictures (including those by Schmalz, whose very name gave definition to a particular brimful sentimentality) talk interestingly about that now 'other' culture. Visitors to the gallery can enjoy the contrast of these paintings, which tend to the didactic, with those of the same period but a different culture. There is a good collection of Louis Buvelot's early Australian work in the gallery. Buvelot was an Australian pioneer of 'plein air' tonal impressionism, allowing the drama and atmosphere of the Australian landscape to speak for itself.

Walter Withers, Rupert Bunny, Hans Heysen, Arthur Boyd, Sir William Dobell, Penleigh Boyd, Grace Cossington Smith, Rah Fizelle and Margaret Preston and

Clifton Pugh are among the many major 20th century Australian painters and sculptors represented in the collection. The gallery recently acquired the a sculpture made by Patricia Piccinini for the Australian Pavilion in the 2003 Venice Biennale, and continues to develop its contemporary collection through purchases, bequests and acquisitive prizes.

Above top: John Gollings, courtesy Bendigo Art Gallery, Bendigo Art Gallery
Above bottom: Agnes Goodsir, Girl with Cigarette C1925 Oil on canvas, Bequest of Mrs Amy E. Bayne, 1945, Collection Bendigo Art Gallery
Opposite top: Edward Hopley, A Primrose from England C1855, Oil on canvas, Gift of Mr. & Mrs. Leonard V. Lansell, 1964, Collection: Bendigo Art Gallery
Opposite bottom: John Gollings, courtesy Bendigo Art Gallery, Bendigo Art Gallery Café

66.

CASTLEMAINE ART GALLERY

The Castlemaine Art Gallery has focussed on Australian art since it was founded in 1913. This policy has, over nearly a century, built an impressive permanent collection. There are representative works by classic artists such as Louis Buvelot, Fred McCubbin, Tom Roberts and Arthur Streeton. Artists of the early 20th century include John Longstaff, Hugh Ramsay, E Phillips Fox, Walter Withers, Rupert Bunny and the Lindsay brothers. Of more recent time are Russell Drysdale, Fred Williams, John Brack, Clifton Pugh, Albert Tucker, Lloyd Rees, Margaret Preston, and Roger Kemp.

In 1990 the gallery began collecting photographs of Australian artists by Australian photographers and holds works by photographers such as Max Dupain, Olive Cotton and May Moore. The collection has benefited from bequests and donations not only of works of art but furniture and decorative objects. A splendid Lipp piano of c.1880 provides a focus for musical events held in the gallery.

Recent enlargement of the building was made possible by the Stoneman benefaction, and there is now an excellent series of well-lit exhibition areas maintaining the character of the original Art Deco interior. Ample curatorial facilities were added at the same time using a Federation government grant.

An extensive reference library is on hand to assist art students and members of the public. The **Gallery Shop** offers postcards and greeting cards, fine art reproductions and books for sale.

The **Castlemaine Historical Museum** is contained in the lower level of the gallery building. It offers a significant collection of historical artworks, photographs, costumes, decorative art objects and artefacts relating to the history of the Mount Alexander district. Much of the Gallery's fine collections of porcelain and gold- and silverware are also on show in the museum area.

Above: Clarice Becket, Boatshed at Beaumaris, oil on cardboard Maud Rowe Bequest, Collection: Castlemaine Art Gallery & Historical Museum
Opposite top: Frederick McCubbin, Winter evening Hawthorn 1886, oil on canvas, gift of JT Tweddle, 1926, Collection: Castlemaine Art Gallery and Historical Museum
Opposite bottom: Clifton Pugh, The Crabcatcher, oil on comp. board, Collection: Castlemaine Art Gallery and Historical Museum

68.

VISUAL ARTISTS

Working artists need to be near a metropolis both for stimulation and to network with their peers. But they also need space, which is at a premium in the city. Proximity to Melbourne, the beauty of the region, and plenty of space for studios, has made the Heart of Victoria an ideal place for artists and craftspeople. Local artists and craftspeople are often represented in local galleries, which also feature work by other major artists with national and international reputations.

There are painters, sculptors working in sandstone, limestone, wood and metal, ceramicists, and printmakers using all the traditional forms including etching and wood engraving. There are furniture makers, glass artists, mural painters, textile artists, potters, jewellery makers, silversmiths and glass blowers, many working in studios that open for visitors.

Some of the artists to look out for include:
- Garry Bish, ceramicist
- Peter D Cole, sculptor
- Jock Clutterbuck, printmaker and sculptor
- Craig Gough, painter
- Anton Hasell, sculptor, & printmaker (Anton created the Federation bells for Melbourne's Federation Square)
- Georgina Hilditch, sculptor
- Ruth Hutchinson
- Andrea Hylands, ceramicist
- Robert Jacks
- Tim Jones, printmaker specialising in woodcuts
- Ben Keith
- Levon Loper, artist in steel
- Glenn Mack, glass artist
- Di Barton, sculptor
- Gemma Nightingale, painter and sculptor
- Fiona Orr, sculptor, with work in the NGV collection
- Russell Petherbridge, a sculptor whose work is popular in the USA
- Greg Preece, painter
- Peter Randall, sculptor and painter
- Wendy Stavrianos, painter
- Noela Stratford, painter
- Dave Lewis, sculptor and furniture maker
- Peter Tyndall, painter who exhibited at the Sydney Biennale
- Allan Wolf Tasker, painter and sculptor
- John Wolseley, painter
- Don Wreford, glass artist represented in the NGV.

Many of these artists are represented by galleries in the region. **Visitor Information Centres** (see separate section) have information about studios that are open to the public, and opening hours.

Opposite top: Ingrida Rocis, 'Rainbow', mixed media, courtesy the artist and Judith Pugh Gallery
Opposite bottom: Ingrida Rocis, 'Bridge', mixed media, courtesy the artist and Judith Pugh Gallery

GALLERIES

Judith Pugh Gallery, **Art at Mount Macedon** (tel 5426 3798), above the village on Mount Macedon Road, is a dealer exhibiting gallery, representing artists from across Australia and the Tasman. Exhibitions feature vibrant modern painting, photography, prints and sculpture. The gallery is in the township's original general store, the only heritage shop to survive the bushfire of 1983. Apart from the charming weatherboard building, however, this gallery could be in any major city. The gallery gallery has a three-weekly exhibition programme, and extensive stock in two stockrooms.

MAD Gallery (tel 5429 1432), in High St, Lancefield, has a large display area with many and diverse works of art. Mad shows contemporary and leading edge fine art – two-dimensional and three-dimensional – from emerging and established artists, from this region and overseas. On the sunny side of High St Lancefield, Victoria, it is one of Lancefield's main points of tourist and artistic interest, with many visitors from Melbourne.

Strawbale Gallery's half-metre-thick walls in Ryan Road, Woodend (tel 5427 1448) make an environmental statement and create a serene environment to display sculpture, paintings, prints and photographs by some of the region's most outstanding artists.

Professional artists honing their skills and those who want to learn attend **Fiona Orr's** sculpture workshops, lectures and conferences in her wonderful studio, an old band hall in George St, Kyneton (tel 5422 2734).

Tin Shed Gallery (tel 5423 2144) in Mollison St, Malmsbury, is a purpose-build exhibition space in a contemporary galvanised iron building. Local artist's works are exhibited and sold, and there are also lectures, courses, tours, art festivals, and music events. **Woodbine Art** (tel 5423 2065), Daylesford Road, Malmsbury specialises in contemporary and botanical art exhibitions of quality artists from Melbourne and the surrounding area.

The **Convent Gallery** (tel 5348 3211) in Daly St, Daylesford is set in an historic building that was built as the local Gold Commissioner's residence and was later a convent and boarding school for nearly 100 years. This lovely building unites history, spirituality, art and culture under one roof. The convent houses seven galleries covering sculpture, fine art, pottery, antiques, jewellery, plus local food and wine. With changing exhibitions every six weeks, there are many unique pieces to choose from. There is an excellent restaurant and comfortable bar, making this the kind of place where you could spend a good part of a day.

Local artists and artisans exhibit painting, prints, ceramics, sculpture and glass as well as fine antique and contemporary furniture in Vincent St, Daylesford, in the excellent **Pantechnicon Gallery** (tel 5348 3500). Internationally-renowned glass artist **Don Wreford's** studio has examples of bright and beautiful bowls, goblets and vases in Albert St, Daylesford (tel 5348 1012).

Heron's Gallery (tel 5472 1030), Templeton St, Castlemaine shows a range of interesting work, and has special exhibitions to coincide with the Castlemaine State Festival.

Allan's Walk Artist Run Space (tel 0428 286 165), Allan's Walk, Bendigo encourages young artists to take risks with their work in all mediums: video, sound, performance and traditional visual arts.

Above: Dora McPhee, 'Autumn Vineyard', 2004, oil on canvas, courtesy the artist and Judith Pugh Gallery
Opposite: Noela Stratford, 'Hanging Rock Profile', 2004, gouache on paper, courtesy the artist and Judith Pugh Gallery

PERFORMING ARTS

A program funded by Arts Victoria tours shows throughout the Heart of Victoria, and audiences are delighted not only by the performances but the wonderful venues in these historic towns. The beautifully restored historic venue the **Capital Theatre** in View St, Bendigo (tel 5441 5344), built in 1873, hosts these regular theatre performances and concerts. The tour also visits the historic **Kyneton Town Hall** and the modern performing arts centre the **Phee Broadway Theatre** (tel 5472 4137) in Castlemaine.

Castlemaine has its own biennial Arts Festival, a fully professional state-funded feast of music theatre and the visual arts. The art galleries in Castlemaine and Bendigo also host frequent concerts by both local and visiting artists. The **Theatre Royal**, built in 1857, in Hargraves St, Castlemaine (tel 5472 1196), is the longest continually operating theatre in mainland Australia offering movies and live shows. Lowland Farm, Mount Macedon, one of the finest modern houses in Australia, is the setting for **Macedon Music** (tel 5426 2414). Jazz, vocal groups and soloists can be heard in a regular chamber music program that combines work from all periods with music by modern Australian composers.

The Mount Theatre in Macedon has a performance program by the Mount Players to suit every taste — it is a professionally equipped theatre, built after the devastating 1983 bushfires destroyed the local hall. On the corner of Honour Avenue at the foot of the mount, the **Church of the Resurrection**, also built post-fire, like many other churches and halls throughout the area plays host to performances by local professional musicians.

Above: Katherine Seppings, Wattle Gulley
Opposite: Pete Walsh, Daylesford

MUSEUMS

Conscious of the importance of history, the people who live in the Heart of Victoria have preserved the history of the district and established many fine museums and heritage displays to share with visitors.

Castlemaine Historical Museum is one of the most professional of these. Housed in an art deco building with the Art Gallery, the museum boasts a significant collection of historical artworks, photographs, costumes, decorative art objects and artefacts relating to the history of the Mount Alexander district.

The **Kyneton Museum** collection includes the building in which the museum is housed. Recognised by the National Trust, the beautiful bluestone building was built in 1855 as a bank of New South Wales branch, to serve the needs of a rapidly-expanding population on its way to the central Victorian goldfields. The exhibits are housed in the former bank manager's residence and stables as well as the servants' kitchen, laundry and cellar, and include a wide range of household items, artworks, photographs, linen, lace, costumes, horse-drawn vehicles and farm machinery.

Daylesford Historical Museum shows a large collection of photographs of the early district and its inhabitants, a display of Djadja Wurrung artefacts, extensive local genealogical material, an excellent display of local memorabilia and artefacts .

The **Making a Nation Exhibition**, in the Old Bendigo Post Office, the **Bendigo RSL Memorial Military Museum**, Pall Mall, and the **Maldon Historical Museum**, offer insights into particular aspects of local history.

The **Golden Dragon Museum** in Bendigo illustrates the story of the region's Chinese miners and settlers who arrived on the goldfields in the 1850s. The museum is home to Sun Loong, the world's longest Imperial Dragon (more than 100 metres long and weighing in 2.5 tonnes).

Specialist industry museums such as the **Bendigo Wine Museum – Balgownie Estate**, Maiden Gully, the **Bendigo Tramways Museum**, **Bendigo Pottery & Interpretive Centre**, Midland Highway, and **Carman's Tunnel Mine**, outside Maldon, offer an opportunity to learn about the systems, machinery, working conditions and traditional skills of crafts and industries important in the 19th and 20th centuries.

Above: Sandy Scheltema, Mount Macedon
Opposite top: Katherine Seppings, Moonlight Flat
Opposite bottom: Geoff Hocking

VISITOR INFORMATION CENTRES

BENDIGO VISITOR INFORMATION CENTRE

Historic Post Office 5167 Pall Mall, Bendigo 3550

Ph +61 3 5444 4445

Freecall 1800 813 153 (within Australia)

Fax +61 3 5444 4447

email tourism@bendigo.vic.gov.au

Web www.bendigotourism.com

CASTLEMAINE VISITOR INFORMATION CENTRE

Historic Market Building 44 Mostyn St, Castlemaine 3450

Ph +61 3 5470 6200

Freecall 1800 171 888 (within Australia only)

Fax +61 3 5471 1746

email visitors@mountalexander.vic.gov.au

MALDON VISITOR INFORMATION CENTRE

93 High St, Maldon 3463

Ph +61 3 5475 2569

Fax +61 3 5475 2007

email maldonvic@mountalexander.vic.gov.au

BLACKWOOD VISITOR INFORMATION CENTRE

21 Martin St, Blackwood 3458

Ph +61 3 5368 6525

Fax +61 3 5368 6834

Email ahaack@optusnet.com.au

DAYLESFORD REGIONAL VISITOR INFORMATION CENTRE

98 Vincent St, Daylesford 3460

Ph +61 3 5321 6123

Fax +61 3 5321 6193

Email visitorinfo@hepburn.vic.gov.au

KYNETON VISITOR INFORMATION CENTRE

High St, Kyneton 3444

Ph/Fax +61 3 5422 6110 or 1800 244 711 (within Australia)

Email vic@macedon-ranges.vic.gov.au

WOODEND VISITOR INFORMATION CENTRE

High St, Woodend 3442

Ph +61 3 5427 2033 or 1800 244 711 (within Australia)

Fax +61 3 5427 3917

Email vic@macedon-ranges.vic.gov.au

Opposite: Alison Pouliot, Mt Alexander

MOUNT MACEDON

Mount Macedon is the shape that appears in the distance and then keeps re-appearing above the horizon on the way north from Melbourne – the long straight slope of 45 degrees signifying the very ancient volcanic core of what was once a spectacular cone.

At first the Macedon Ranges look like a simple row of forested hills. A closer view exposes their complexity. When seen from New Gisborne or Macedon village, the ranges disclose valleys going deep into the heart, with the deepest profile near the western end of the ridge. From this view the extent of the original crater or volcanic caldera can be estimated - from left of the summit (where the Memorial Cross stands) to the crest of Mount Towrong. This means that the entire village of Mount Macedon and the great estate gardens are found within the actual crater. The intersection of Devonshire Lane and the mountain road is about the centre of the ancient cone.

The bump on the northern profile is Camel's Hump, a recent eruption (a mere 6 million years ago) on the rim of the crater, and dating from about the same time as Hanging Rock 2.5 km north. The date of the original volcano is buried in the mists of time – between 360 and 370 million years ago.

It was the good soils and higher rainfall up the slopes that led the government of the day to delay any settlement until they had an idea how to make best use of these conditions. Timber-getting was the first activity and has remained a major factor in the local area until the present time. The Black Forest Mill still operates two km south of Woodend, the last of many mills originally in the area.

Above: Daryl Chibnall, Mt Macedon Caldera
Opposite top: Sandy Scheltema, Mt Macedon
Opposite bottom: Katherine Seppings, Mt Macedon

The township of Mount Macedon has long been a separate community from Macedon village. After a time as a woodcutters' settlement in the later 1850s and 1860s, it grew with the establishment of large hill estates from 1872 onward. The passing of government legislation in that year encouraged landowners to buy seven-acre blocks for planting cool climate softwoods. It was believed the deep loam soils and high moisture levels would encourage quick growth in conifers for building-timber and deciduous furniture-woods, which would be a valuable resource.

The actual outcome was that wealthy people bought the blocks and, where they could, assembled two or three together to make estates of fourteen and twenty-one acres. This was big enough to enable the establishment of a large garden with the necessary outhouses and land for vegetables and dairy cows which were needed if you were planning to run a sizeable estate. Forty such large estates were developed on the mountain after the example of the hill stations of British India.

In time the gardens of Mount Macedon became famous for their beauty and variety. Nurseries were established nearby and social networks built up for this rarefied community. Those who lived on the mountain full time were generally workers on these grand estates; they kept things in order until the warmer months when owners would come up from Melbourne. Large convoys would arrive with extended families, household servants, cows and more servants. Over summer, house parties were held, tennis courts rang with cries, and croquet lawns sighed with muttered curses. Picnics were held along the forest tracks and life sparkled.

Today not many of the original estates remain intact; changing times and bushfires having taken their toll, but enough of the grand houses and gardens remain to give a picture of life as it was lived. There are a number of newer gardens and notable dwellings, but the needs of a large establishment are now different. Gardeners live away from the job, and servants are not often found in-house. There are probably another forty newer gardens or estates today, and there is always a queue at the gate for garden-open days. Australia's Open Garden Scheme brings new visitors each year, and publicises the special character and history of the gardens of Mount Macedon. Worthy charities have their day in some gardens, and the local Horticultural Society organises visits and conferences on an ever-widening array of related subjects.

Above: Katherine Seppings, Mt Macedon
Opposite: Sandy Scheltema, Mt Macedon

The old village of Mount Macedon was on a steep section of the main road about halfway up the rise; one original store remains, now the Judith Pugh Gallery. All the rest were swept away on Ash Wednesday 1983, along with more than half the houses.

The **Mountain Inn** survives and is a popular local watering-hole, with a busy restaurant. **Trading Post** is the local store and post office, doing a busy trade in coffee and lunches. Three nurseries are found here, each with a different range on offer: **Dicksonia** specialises in rare plants, **Florilegia** in perrenials, while **Newmans** has a general range of shrubs, trees and bedding plants.

Today this is a growing community, with a lively primary school and bright outlook. The tiny Presbyterian Church, rebuilt, is now the Uniting Church. Two local Anglican churches were destroyed and opportunity was taken to combine to build one new building, the Church of the Resurrection – just at the foot of the mountain. This particularly beautiful church, designed by Canberra architect Bryan Dowling, is lined with blackwood timbers milled from trees burnt in the fire. The east window commemorates those who perished in the 1983 fire - they are depicted with the suffering Christ in the Daille de Verre glass made by Leonard French.

The mountain road leads up the hill with typical large properties on either side. Most of the larger estates are concealed from view by hedges or shrubberies. Enough of the older gardens have survived the fire to give an idea of the size and maturity of those lost. Many of the exotic species are fifty metres high or more, after 130 years of growth.

The road continues to the far northern summit of the ridge where a left turn leads to the Memorial Cross on the western end of the mountain. There are endless views over the Melbourne plain, Black Forest and tablelands to the west. The **Top of the Range** kiosk provides food and drink to mountaineers. Camels Hump, near the main road, is the highest part of the mountain, and is worth the climb for the view north.

On the main road down the mountain, **Blue Ridge Inn** is seen on the second hairpin turn. This excellent new hostelry has probably the best view in the district. Soon after, the right-hand turn into Straw's Lane is recommended for a quick trip with glorious glimpses into the valley and to Hanging Rock.

Opposite: Sandy Scheltema

MACEDON

Macedon grew with the wood-cutters during the 1850s, and has always been a substantial village. It was surveyed in 1860 and officially named Macedon. The building of the railway brought growth to the town, and the contractors Cornish & Bruce of Castlemaine actually built a church for the fettlers in Macedon in 1860. The village continued as a workingman's town until quite recently. It has always had a life distinctive from Mount Macedon.

There was a considerable town centre in Victoria St with about twenty shops, mostly timber-built with shady verandas to the street. But the fire of Ash Wednesday 1983 removed almost the entire village, on a cataclysmic night when all who could sheltered in the Macedon Family Hotel and miraculously survived. Others were lucky to escape with their lives, all houses and property going into the flames.

Today the community is active and growing, and many new houses continue to appear in the area. There are a half dozen shops and businesses in Victoria St, the Macedon Family Hotel prospers with a good dining room, there is a wine bar, a police station, two churches and primary school. Most of the structures are new. There is a new Memorial Hall and a theatre, The Mountview, replacing one lost in the fire. The Mount Players deliver a year-round program to the highest standard of amateur theatre.

Nurseries have long played a part in Macedon. Today there are two, the Mountside who are landscape suppliers, and Macedon Nursery on the site

of the former state forestry nursery. The latter has a vast range of natives and exotics, specialising in cool climate plants.

Macedon Cemetery is a quiet enclave, well-cared for and the subject of a recent book. It has a number of fascinating graves, including those of Lord and Lady Casey, Ellis Rowan (artist) and a beautiful sculptured female form on the Mathieson tomb.

Above: Daryl Chibnall, Macedon
Opposite: Joe Mortelliti, Prendergast Lane, Cobaw Range

86.

WOODEND

Woodend is an old town on the Dividing Range. It was a wayside stop on the road to the goldfields and remains a rest stop today. It has expanded gently in the last decade, many commuters choosing a country lifestyle. The town has gone through a renewal experience since the freeway bypassed the main street in 2000. Regular rail services link to the metropolis, so commuting is easy. The schools and churches, community sporting groups and other activities are lively and growing.

A pretty High St crosses the railway and runs down the slope towards Woodend Creek. The **Town Clocktower** stands in the centre of a broad avenue near the former **Post Office**. Almost opposite is the old **Institute Building**, very grand but no longer a community centre, having become legal offices.

Every kind of eatery is available, from fast food to more relaxed and considered eating. Good fare is available at **Holgate's Brewhouse** (Keating's Hotel), **Café Colenso**, **Wildwood**, **Literary Latte** and other cafés across the road. **Burke's Bakery** is at the end of the shopping strip. The **Fruit Market** has consistently delicious products, especially cool climate fruit and vegetables from the local region, including apple and pear juices from Harcourt. The well-stocked **Wineshop** has a representative range of wines from the Macedon Ranges and Central Highlands.

Woodend is a good jumping-off point for the Macedon Ranges, Hanging Rock and the region towards Trentham and Daylesford. Excellent boutique hotels include **Campaspe House** and **The Bentinck**, and there are many B & Bs to choose from.

Above: Geoff Hocking
Opposite top: Daryl Chibnall, Woodend
Opposite bottom: Daryl Chibnall, Woodend

HANGING ROCK

Hanging Rock sits in the middle of a beautiful valley immediately north of Mount Macedon. At this point the Macedon Ranges run east-west. Beyond the rock to the north is a small range of hills, then the larger wooded range of the Cobaws.

Hanging Rock is best reached from Woodend along Romsey Road, or by continuing on the mountain road over Mount Macedon and turning down Straw's Lane — there are glorious views into the valley.

Hanging Rock was always a local curiosity, popular with picnickers at least since the 1860s. The owner of a small hotel nearby established regular sporting events and laid down a racecourse on the western side of the Rock. Picnic race meetings were held, attracting crowds in thousands. This racecourse has since migrated to the opposite side of the rock, and race horses now circle a small lake set amidst twisty gums. Regular sporting and social events are held here each year, but the chief attraction is climbing the rock itself. A restaurant, souvenir shop and small museum exhibit make the visiting charge worthwhile.

Hanging Rock is a mamelon, a conical hill formed by the extrusion of highly viscous lava, called soda-trachyte. The other known occurrences of such rock are in Abyssinia and Norway. This rather soft lava has eroded over the last six million years to form vertical pinnacles that are the striking element of its appearance.

There are two other similar formations in the vicinity; **Camel's Hump** on the northern rim of Mount Macedon 2.5 km directly south (300 metres higher up), and **Brock's Monument** 8 km east, smaller than the other two.

Remarkable as Hanging Rock is, such fame could not have come its way without the help of two interdependent occurrences. The first was the publication of Joan Lindsay's popular book *Picnic at Hanging Rock*, which sets up the entirely fictitious story of a St Valentine's Day school picnic where several girls and a teacher go missing, never to be found. The Peter Weir movie of the same name immortalised the story and gave it a world-wide audience. In both accounts the air of mystery that inhabits the Rock is the prime element of the story.

Among the stone pinnacles or even near the base of the mount, the atmosphere is undeniable, but hard to define. Many will testify that *something is going on…* It was a site of significance to local Kooris, but it is not known in what way. A group of Aboriginal teenagers visiting some years ago from the Northern Territory, ascended the rock happily before one of them sensed something; when he shared his feelings, the others too became alarmed and hurried down. They refused to return.

Hanging Rock is surrounded by pretty lanes on all sides, thick with small trees and glimpses of surrounding hills. Two km north is **Newham** — with a village store, school, community hall and a scatter of houses. On the hillside above the village, **Hanging Rock Winery** is a well-known wine-tasting facility with lodges for short-stay B & B. The view over the vineyard towards Hanging Rock and Mount Macedon is superb. East of here are other top-ranking wineries, including **Rochford**, **Glen Erin** and **Cope Williams** (the two latter with restaurant and conference facilities). Beyond Deep Creek are **Cleveland**, **Portree** and **Mount William**.

Opposite top: Daryl Chibnall, Hanging Rock
Opposite bottom: Daryl Chibnall, Hanging Rock

LANCEFIELD & ROMSEY

The small rural towns of Romsey and Lancefield lie east of the Macedon Ranges and within view of the pretty volcanic range that ends with Mt William. Neither has a population of more than 1500, but each town has a grand layout with a boulevard at the centre, expecting a greater public - which they might have some day, as Melbourne's fringes expand.

These towns feel very different from the central goldfields towns, because they are primarily rural. They have always been more focused on potatoes and dairying than on gold. The railway came in 1881 and went in the 1950s. Lancefield Station is now a plant nursery and traces of rail-bed and old timber-trestle bridges are still seen near the main road.

They are wide-open towns with fine views of surrounding hills at all times. The streets are lined with great English elms and oaks, and stands of conifers denote parks and larger homes. And the countryside between them shows fine stands of exotic trees, with few native species left to remind us we have not quite left Australia.

Lancefield was built at the edge of Deep Creek, which originates on the north-east flank of Macedon Ranges. Floods in the late 1840s and early 1850s caused the town to move up the slope – but not too far. Deep Creek soon develops its own gorge to flow south across Melbourne's basalt plain between 50 and 100 metres below the general level, arriving at Footscray disguised as the Maribyrnong River.

Lancefield's notable buildings include **The Institute (Community Centre)**, two grand former banks, a matching courthouse and post office (in different streets), a good butcher, an art gallery that serves good coffee and a couple of pretty churches. A gentle stroll or drive along residential streets

brings the reward of delightful old houses and cottages. The most significant building is an immense former hotel – **The Macedonia** – now part wine-bar, part antique shop and restaurant.

There is a Farmers' Market in the grassy centre of the main street the fourth Saturday of each month.

Burke & Wills Expedition camped at Lancefield overnight in 1860, and are said to have left without paying for goods or liquor. The road northwest to **Baynton** is still called **Burke & Wills Track**. It leads through forested hill country towards Heathcote and Mia Mia. Near Baynton is **Knight's Granite Hills Winery**, which has an excellent range of wines and spectacular views across dancing hills to distant Mount Alexander.

Above: Joe Mortelliti, Cobaw Range
Opposite top: Joseph Kinsela, Lancefield
Opposite bottom: Joseph Kinsela, Lancefield

KYNETON

The town of Kyneton is built on a gentle westward slope and sits within a right angle described by the Campaspe River. The town, unlike most settlements in this region, predates the goldrush. It grew from the earliest run, **Carlsruhe**, taken up in 1837 by C H Ebden. As other settlers moved into the area, the station was broken up. In late 1839 James Donnithorne took over the lease of the northern part and named it **St Agne**s. A station complex was built on the west bank of the Campaspe where **Rock House** stands today, next to the present homestead called St Agnes.

When the neighbouring station built cottages on their side of the river, the place began to resemble a village. A visit from the Lieutenant-Governor Charles Latrobe in 1848 led to the official decision to gazette the town. By 1850 Kyneton was a growing rural centre, just in time to play host to the human flood after the discovery of gold at Mount Alexander the next year.

Because it took shape while Victoria was still a quietly promising colony based upon pastoral and maritime pursuits, Kyneton has a pre-Victorian character. The discovery of gold brought a sudden rush of new entrepreneurs, architects and builders to the region and Victorian style suddenly became the thing. It is a bluestone town: a high proportion of the town's 19th century buildings are of squared or rock-faced basalt, and brick does not make an appearance until quite late in the century. The earliest house remaining is the **Old Rectory** (now a B & B with a Paul Bangay garden), on the corner of Piper and Ebden Sts, built in 1851.

Above: Daryl Chibnall, Kyneton
Opposite: Daryl Chibnall, Kyneton

PIPER ST

Piper St was the commercial centre during the 1850s and has not changed much. The whole street is a remarkable survival and still has original bluestone paving, kerbs and gutters, and gaslights. Today there are interesting activities - antique shops, a restaurant or two, a bakery and some fine old pubs. There are rows of substantial shop-fronts, and it comes alive at weekends and public holidays.

The first Bank of NSW (1855) now serves as the **Museum** (corner of Powlett St) and has an excellent display of 19th century interiors with costumed figures, and an original squatter's house and stables. Two original flour mills remain: **Willis Brothers Mill** (1862) on the corner of Ebden St with machinery intact and still making flour, the other now known as the **Old Butter Factory** (corner Wedge St) which contains **Meskills Woolstore**, selling pure spun wool, knitting yarn and knitwear.

Halfway along is **St Paul's Park** with a view up the hill to the church; a **Farmers' Market** is held in St Paul's Park the second Saturday of each month. **St Paul's Anglican Church** is best approached from Powlett St where the pathway rises up the hill with dramatic views of the handsome tower through mature stands of Atlas Cedars and a large Lebanon Cedar close to the church.

On the corner of Piper and Mollison Sts, **Zetland Lodge** (Freemasons) is a rather plain redbrick building – however the Lodge Room inside is a well-preserved example of ritual furnishing used in Freemasonry. The highlight is the decorative scheme, including scenes of the Nile and the Pyramids beautifully painted in deep perspective by **Thomas Fisher Levick** during the 1920s. Levick was art-teacher at Kyneton and Castlemaine, and a founding spirit of the Melbourne Workingman's College (later the Prahran Institute of Technology and now Swinburne University).

Ellim Eek, at the intersection of Mollison and Mair Sts, was built in 1890, then purchased for a surgery by a local doctor who added Oriental terracotta ornament on the parapet, coloured ceramic panels, and an unusual veranda.

Above left: Katherine Seppings, College House, Piper St, Kyneton
Above right: Joseph Kinsela, Zetland Freemasons Lodge
Opposite: Martin Hurley, Piper St Kyneton

MOLLISON & HIGH STS

The event that took commerce away from this end of town was the arrival of the railway from Melbourne on the other side of the Campaspe River. Trains began running in 1861 and the town grew out to meet them. The business centre has ever since been along Mollison St and its intersection with High St (the road to Melbourne). The large bluestone station is a rare example of an 1860s station with all its appurtenances intact – level-crossing gates, signals and goods sheds.

The present commercial centre in Mollison and High Sts dates mainly from the 1860s to 1910s, with some older buildings in High St. It is an impressive townscape with a few major gaps on street corners – thanks to premises associated with selling or servicing the motor vehicle.

The magnificence of the old bank buildings testifies to the financial strength of Kyneton through the late 19th century – the **National Bank** still operates from original premises designed in 1877 by Leonard Terry. Opposite is the art nouveau former **Bank of NSW** – the date 1817 refers to the founding of the banking company, not the building which is nearly 100 years younger.

Kyneton Post Office is a superb civic building, mainly in Italianate style with an open arcade on cast-iron columns – but the clock tower is more Gothic with a clock face on the diagonal and spiky pyramid roof. The **Town Hall** was begun in the late 19th century and extended in the same handsome style when the hall was built as a memorial to WWI. The excellent hall functions as the town's civic centre and hosts musical and theatrical presentations travelling Victoria's major country centres.

Past the traffic lights is the old **Mechanics Institute**, perhaps the finest original building in town. The temple-like structure has a two-storied central façade with Doric pilasters and pediment; the symmetrical wings are single-storied. Inside is an atmospheric hall with stage and gallery, now used as a

community centre. The town library on the right-hand side is due to be renovated and extended, while the rest of the building is to be restored; all this with the help of Kyneton Bowling Club who have a long-term lease of the rear of the building.

Further along **Mollison St** are interesting premises from every decade of the 19th and early 20th Centuries, commercial and domestic, and the **Botanic Gardens**. The two principal secondary schools in High St are worth a look. The **State Secondary College** has an elegant 1920s classical front, and almost opposite are the spectacular redbrick Gothic convent buildings now used as the **Sacred Heart Catholic Secondary College**.

The old churches of Kyneton are best seen from Ebden St – all are bluestone and have lots of character, but particularly the **Holy Rosary Roman Catholic Church**. Many interesting and beautiful old houses are to be found in the streets of Kyneton, and some characteristic timber or brick houses of the inter-war period with bow-fronted windows. One superb bluestone structure is the former **Kyneton Hospital**, well proportioned and with an excellent cast-iron veranda – at the western end of Simpson St.

Above: Katherine Seppings, Cenotaph & Mechanics Institute
Opposite top: Katherine Seppings, Shamrock Hotel and Post Office
Opposite bottom: Katherine Seppings, Piper St, Kyneton

AROUND KYNETON

The Calder Freeway after Woodend gives clear vistas towards Hanging Rock and the northern face of Mount Macedon. Then **Jim-Jim Hill** at Newham shows its complex volcanic profile and the **Cobaw Range** further north, grows distant as the road follows the valley of the Campaspe River past **Carlsruhe** towards Kyneton.

The old bluestone mill near the river is the **De Graves Mill**, and nearby stands **Skellsmergh Hall**, a grand bluestone house with wide verandas. Not used since the 19th century, the mill is an industrial relic and gives witness to the importance of the district as a grain bowl for the colony. There were six flour mills in Kyneton during the first twenty years of the goldrush.

Huge mills were erected in the 1850s when demand for grain was heightened by the sudden influx of population on the goldfields. The country was good for wheat and grain because the virgin loamy soils produced well, but it soon became clear the lighter soils of the north and west of the state were easier to plough and less boggy for reaping. Cereal crops quickly followed the opening up of these agricultural broad acres, leaving the central highlands to pastoral pursuits, traditionally cattle and sheep with a drift towards exotic stock such as deer and alpacas.

Another long-time trend is to vineyards, with larger acreages under vines each year. The short cool summers and late ripening makes for autumn vintages and light wines. Wineries are located north and east of Kyneton. Locally made olive oil has come on the market as **Kyneton Olive Oil**. The olive groves are near **Barfold** on the Heathcote Road. Table olives are also available.

The remnant gold town of **Lauriston**, to the southwest, has two reservoirs on the Coliban River, where fishing and boating are permitted in season – inquiries should be made first.

Kyneton Mineral Springs are 2 km west of the town on the Old Calder Highway (Piper St). There are trees, picnic facilities, a hand pump for mineral water and a period rotunda. Kupper's Roadhouse Café is opposite.

Above top: Joe Mortelliti, Redesdale
Above bottom: Joseph Kinsela, St Agnes
Opposite: Geoff Hocking, Sutton Grange towards Macedon

Harcourt Road goes north past the **Black Hill Reserve**, a 60 hectare conservation area for native flora and fauna. Black Hill is an ancient granite outcrop, very crumbly, surrounded by a sea of basalt flow. Although not of any great height it is locally prominent like its neighbour Bald Hill, and on an early map of 1859 is marked as on Major Mitchell's return journey from Mount Macedon in 1836. Every sort of eucalyptus and acacia of the region is found on the reserve and a wealth of native flowers and orchids (see the excellent booklet by Lois Prictor, A Walk Through Black Hill Reserve – Flora and Fauna of Central Victoria - 1987).

Further along the road is **Redesdale** where an atmospheric old pub is the social centre. After the handsome **St Laurence's Church** the road descends suddenly into the narrow valley of the Campaspe River and crosses on a wonderful iron-framed **double bridge** – one structure for each lane of traffic.

The village of **Mia Mia** lies at the intersection of this road with the **Burke & Wills Track** from Lancefield. Wineries in this area are **Eppalock Ridge** near Redesdale and the famous **Knight's Granite Hills** towards Lancefield (Baynton).

Lake Eppalock is seven km north of Mia Mia. There is fishing and boating in season – inquiries should be made first.

East of Kyneton, **Pipers Creek** and **Pastoria** may be included in a 20 km round trip or as a way of visiting the **O'Shea & Murphy Vineyard** at Rosebery Hill, and **Cobaw Ridge** at Pastoria East. There's the delightful scenery of the downlands towards the Campaspe valley, with Mount Alexander distant to the north-west and the Cobaws to the east.

There are a number of B & Bs in town and countryside, including **Paramoor Farm Country Retreat** (with a real barn & old farrier's shop) in Three Chain Road, **Riverlodge Retreat** (in Carlsruhe village) and **Pond Cottage** at Glengrove Farm in Chases Lane, Pipers Creek

Above left: Katherine Seppings, St Agnes, Kyneton
Above right: Katherine Seppings, Kyneton
Opposite: Chris Kirwan, Bayton

Gary Chapman, Muckleford

104.

MALMSBURY

Beside the freeway, before Malmsbury, there is another huge, bluestone Mill, built in a style already old at the time – resembling mid-18th century mills in the north of England. This is one dates from 1856-7 and operated as a flour mill.

The entry to Malmsbury shows the first of the dramatic changes to the countryside common from here to Bendigo. The town sits in a deep valley where the road crosses the Coliban River. This is also the entry to the Victorian goldfields. The dam wall of the Coliban Reservoir is directly up-river from the railway viaduct and can be seen from trains approaching the station. The channel flows under the Calder Highway just past the Coliban River bridge. Water is reticulated from here via Harcourt to Bendigo, Castlemaine and Maldon.

Malmsbury expected to be bigger than it is, and there are several pieces of evidence that give this away. The most obvious is the layout of the town, which is designed to accommodate a large population. The grand avenue through the centre, Mollison St, is used as actual roadway for less than half its width and shops are only found on the southern side.

The **Anglican Church** was built to seat over 450 people, commensurate with a parish three times the size of the town in its heyday. This fine old bluestone church can be seen from the main road on the north-eastern slope near the town-centre. The church, designed in 1861 by Melbourne architects Purchas and Swyer, comprises a double nave divided by arches. The unusual façade has a star-shaped rose window and an octagonal belfry.

Malmsbury has long been a favourite stopping place for eating and sightseeing. There are two art galleries: **Tin Shed** in the midst of the shops, and the **Old Mill** near the top of the next rise. **Malmsbury Bakery** is ideal for a relaxed lunch and has a wine bar open late in the week; there's a greater choice of venues at weekends.

A number of pretty buildings can be seen from the road, most notably a grand homestead with a vineyard on the north-west slope. The road to Daylesford crosses the Bendigo line on a handsome arched bridge, giving a view of the bluestone railway station (1861). Further along the Daylesford Rd there are several vineyards worth visiting. On Zig-Zag Rd, North Drummond, are Basalt Ridge and Zig-Zag Wineries. Further toward Glenlyon is the Big Shed Winery at Denver, and just on the western side of Glenlyon village is Laura Glen Estate.

Above: Katherine Seppings, Malmsbury
Opposite top: Daryl Chibnall, Malmsbury
Opposite bottom: Katherine Seppings, Malmsbury

Gary Chapman, Maldon

TARADALE

The Calder continues toward Taradale along a basalt-flow ridge with views over forested ranges to the west. The ridge has been mined by shafts and tunnels; huge mullock heaps are seen on either side of the road. The railway to Bendigo runs on the eastern edge of the same ridge, keeping in view the Coliban River.

As the road descends into the village of Taradale, the magnificent railway bridge can be seen to the right. It consists of alternate stone and steel piers supporting a steel-trestle deck. The bridge is forty metres above Back Creek, and was for many years the highest on the Victorian railway system. The road that leads beside the creek should be followed as far as the base of these great piers, so that their true scale can be appreciated. There is also a path through the Fairy Dell (elm thicket) along the creek with special interest for children young and old.

Taradale has lost many houses over the last century. Most were timber-built and were either transported to another place or simply crumbled away. The former **Methodist Chapel** is now a charming residence and behind is the old courthouse. On a hill to the east stands **Holy Trinity Church** with a Wagnerian façade (1858); along with the local school and community hall, it is one of the only facilities in town still in use.

Taradale Mineral Springs Reserve is on the creek behind the Fire Station. There are several local walking trails around the town. Towards Elphinstone the water-channel from Malmsbury Reservoir passes under the road, and can be seen winding around the hillsides on its contour.

Above: Gary Chapman, Taradale Viaduct
Opposite: Daryl Chibnal, Taradale

CASTLEMAINE

Castlemaine is the golden town in the Heart of Victoria. It grew up at the epicentre of the Mount Alexander goldfield – at the intersection of several valleys and the natural access from one mining area to the next. Castlemaine, with its neighbouring town Maldon, gives us the best insight today into the hectic era of the goldrush 150 years ago.

From 1852 this area was thronged with miners, turning over the top six metres of soil in what was to prove the richest shallow alluvial goldfield ever discovered. Known as Forest Creek Diggings, the population of this region soon exceeded that of Melbourne. Within months the government had established a camp, with police contingent, courthouse and gold commissioner. The site for a town was surveyed and building allotments sold, and a more permanent settlement began to appear. Many of the buildings that characterise Castlemaine and the neighbouring villages were built during the next twenty years, for after that time the population dwindled from a peak of 30,000 to about 7,500 as the surface gold gave out.

The Maldon Rush (west of Forest Creek) responded well to deep lead mining, with activity continuing well into the 20th century, but Castlemaine has remained the commercial and administrative heart of the district. The current population of Mount Alexander Shire is about 16,000, including Maldon.

Local townships are so closely integrated with the remaining evidence of mining – from the small-scale panning of the first decade to the last days of widespread sluicing - that the state government has included much of the area in the **Castlemaine Diggings Heritage Park**. Visitors can now travel back in time by participating in guided town walks, self-guided local walks, or by following the published guidebook Discovering the Mount Alexander Diggings which interprets the mining activity on each site.

Although first and foremost a town of the goldrush, Castlemaine is also the paradigm of every country town in Australia. It is fine and picturesque, but rather disorganised. Some of its best treasures and historic delights have to be sought out, or have since been transformed to serve some other function. Street vistas are often cluttered with power poles and cables, and many old commercial buildings have been disfigured by insensitive awnings or bright paintwork. Nonetheless, this is a substantial regional town with a considerable commercial and cultural heart, and a lasting sense of its own history. There is much to be discovered.

Above: Katherine Seppings, Theatre Royal
Opposite: Daryl Chibnall, Castlemaine

TOWN CENTRE

The best-known building in Castlemaine is the **Old Market Hall** in Mostyn St, dating from 1861 and designed by W Downe in the Palladian manner, with a grand Doric portico attended by domed corner-pavilions. The spacious interior is roofed by arched beams of laminated timbers, which look somehow modern. It was used as a market for more than 100 years, now serving as the town's principal **Visitor Information Centre**.

The **Theatre Royal** in Hargraves St is obviously the local cinema - its Art Deco front gives it away. But it is more than that – it's the oldest continuously-operating theatre on the Australian mainland. Its doors opened in 1855 when the surrounds were a tent town set in a sea of muddy diggings. The famous temptress Lola Montez performed her dances here in 1857, and many other notables 'walked the boards', including Madam Melba.

Photographs from the early days show the theatre changing face, but not form, at least three times. The late 19th century interior had a horseshoe balcony, deep arched proscenium and pretty, flowery ceiling. Repairs after fifty years removed many of the best features, but its present appearance is a complete make-over from 1937. Movies can still be enjoyed, and live concerts or cabaret in season. Coffee, ice-creams and cocktails are served in the foyer.

Nearby are antique shops, including the remarkable **Restorers Barn**, which is an extraordinary treasure trove and much bigger than it appears from the street, and the excellent **Stoneman's Bookshop**. **Chapman's Fine Framing & Photography** has outstanding panoramic photos of the region on sale. Art galleries are found in Hargraves St (**Heron's**), Campbell St (**Nunan**) and Templeton St (**Falkner**).

The finest group of buildings in Castlemaine forms the administrative heart in Lyttleton St. This comprises a row stretching from the 1877 classical **Courthouse** westwards to the **Post Office**. The 1889 **School of Mines** façade displays subtle classical motifs and confidence of proportion - the hallmark of Bendigo architect WC Vahland. However, Castlemaine's great centrepiece is the **Post Office** of 1875; designed by JJ Clark in the office of the Government Architect, it is a classical building of such power and discretion that it ranks with the best in the state. The superb clock tower has a clock by Gaunt of Melbourne and a bell cast at Horwood's Albion Foundry. Next to it is the old Drill Hall, solidly timber-built and handsome. The **Town Hall** (1898) has been improved by recent restoration, but in such distinguished company its jolly architectural efforts seem to be less than serious, resembling an Edwardian mantelpiece.

Across the street are two buildings familiar to viewers of Blue Heelers: the former **Imperial Hotel** with its extravagant balconies and mansard roof, and the cool blonde-brick police station.

Above: Geoff Hocking, Old Market Hall, Castlemaine
Opposite: Daryl Chibnall, Castlemaine

Victory Park is on the corner of Barker and Mostyn Sts and this houses a remarkable structure: a large bluestone drinking fountain. In fact, under the canopy there a drinking fountain for people, nearest the corner there's a trough for horses and at the sides there are lower troughs for dogs! The design was by Thomas Fisher Levick, an art-teacher at Kyneton and Castlemaine and a man of unusual perception. The fountain was built as a memorial to Sir James Patterson, who went from Mayor of Chewton to State Premier (1893-4).

Several handsome bank buildings face the park, two of them – the **National Bank** and **ANZ** – are still used for their intended purpose. Perhaps the best is the former **Oriental Bank Chambers** (now solicitors) in Barker St, a redbrick two-storied structure with exquisite proportions and Georgian detail. In Hargraves St, the old **Savings Bank** of 1855, a Mannerist façade - that is, classical with attitude - with an extraordinary gorgon's head as keystone of the arched doorway. This was designed by local architects Poeppel and Burgoyne. After 1921 the building was used as a police station for fifty years. The over-sized head would have made an impression upon all who entered.

The town's **Art Gallery & Historical Museum** is in Lyttleton St. It has a handsome art deco façade designed by Percy Meldrum with sculptured figures in the panel above the entrance. The gallery has specialised in Australian art and artefacts since its formation in 1913; represented are famous names such as Tom Roberts, McCubbin, Walter Withers, and Rupert

Bunny. In the basement there's a well-selected display of historical items, photographs and locally made silverware, along with the gallery's collection of fine china.

Above top: Katherine Seppings, Castlemaine Gardens
Above Bottom: Katherine Seppings, Mostyn St
Opposite: Katherine Seppings, Barker St

116.

AROUND TOWN

Other monuments are found about the town. Castlemaine loved to honour its famous citizens and champions. The largest memorial by far is the splendid **Burke & Wills Monument** looking down Mostyn St from the east. It is a huge obelisk made of grey Harcourt granite in 1862 to honour the ill-fated explorers, especially Robert O'Hara Burke, stationed here as Police Inspector for four years.

A large brick warehouse behind a petrol station in Barker St has superb double arched windows and emphatic quoins - no ordinary shed, this. A steam flour-mill in 1857, it has been the site of a portfolio of industrial activities, becoming **Cornish & Bruce's Railway Foundry** in 1859. A quartz-crushing plant was added at the rear for miners to bring in their raw materials, and the **Castlemaine Distillery** also shared premises for a time. The beautiful tapered chimney stack was demolished early in the 20th century. Another well-known foundry, **Thomson's**, survives from 1875, the oldest engineering company still operating in Victoria (Parker St).

The **Railway Station**, marshalling yards and train sheds all date from the earliest years of the Bendigo Line (1861-2). Nearby bridges and underpasses are similarly sturdy brick structures in original state. The **Central Goldfields Railway**, with steam trains operating from Maldon, will soon be arriving at Castlemaine - after thirty years absence.

Residential streets display interesting houses of all styles and sizes, from the humble to the grand. Quite a few of the best serve as B & Bs, and one good Victorian-classical pair is now the **Campbell Street Motor Lodge**.

BUDA

The best known house in Castlemaine is Buda, a rather grand place with a hint of Middle Europe. The original bungalow was built in 1861 for a retired Indian Army colonel. When the celebrated gold- and silver-smith Ernest Leviny bought it two years later, he named the house for his hometown, Buda (the sister town to Pest, in Hungary). Leviny's daughters left their home and contents to the townspeople. It is now administered by a trust, and open to the public. Two acres of sunken garden provide a model of floral display in a hot dry climate; huge hedges and shady walks invite visitors to linger.

Above: Katherine Seppings, Buda
Opposite top: Katherine Seppings, Town Hall
Opposite bottom: Katherine Seppings, Town Hall

CASTLEMAINE GOTHIC

On the heights around the town-centre there are a number of identifiably Gothic gables and spires of local churches, most of which appear in the earliest pictures. The Gothic style also extends to the earliest public schools and many local houses - here and there a steep gable or frilly barge-board peering across a garden fence.

Castlemaine has the usual bounty of churches and chapels dating from the gold rush – not that miners were noticeably more religious than their contemporaries, but there was considerable effort from each Christian denomination to get structures in place in such an unruly society.

In Lyttleton St, up from the art gallery, there are two remarkable church buildings. The former **Congregational Church** (now Presbyterian) built in 1861 has astonishing presence – from its triple gables framed by immense pinnacles to the windows filling most available wall space. Its Early English Gothic style is presented as High-Victorian drama, in cheerful redbrick with cement dressings.

Across the street is the original **Presbyterian Church** (now Uniting), one of the finest Arts & Crafts compositions in Australia, designed in 1894 by C D Figgis of Ballarat. The architect described it as an adaptation of Florentine or Paduan Gothic, but he may be referring to the theatrical element. It certainly reaches for the skies. Well–observed English and French Gothic details pile up to great height. The main entrance is a spectacularly large arch containing a pair of doors with chequered tiles above. On the gable the cross of St Andrew is set on the diagonal. The bell cast roof of the octagonal spire echoes the west window. The rest of the building is a model of quality brick and cement-stucco.

The two main public primary schools of the town are also essays in the Gothic style, built in the 1870s. **Castlemaine north Primary School** (Barker St) is an impressive example of 19th century School Gothic. There are lots of windows and high roofs for good ventilation. A solid tower with a slender spire marks the main entrance, vertical counterpoint to a long horizontal structure in bi-coloured brickwork.

Another outstanding Gothic example is **St Mary's Roman Catholic Church** in Templeton St. The site gives it prominence enough, but being painted white gives it even more. St Mary's was designed by the locals Poeppel and Burgoyne, yet is different from their work across town at Christ Church. More Germanic, it has a beautiful interior with good stained glass.

Above: Joseph Kinsela, Castlemaine
Opposite left: Joseph Kinsela, Castlemaine Uniting Church
Opposite right: Joseph Kinsela, Castlemaine

120.

CHRIST CHURCH & AGITATION HILL

Christ Church in Mostyn St (Anglican) was the first to get going, with foundations laid two years after regular services began in 1852. It is the most traditional example of the Gothic style in town, built of local sandstone-shale, with carved leaves and faces on the window-hoods and a big rose-window above the west door. Although it looks very English, the church was designed by Poeppel, a local architect who, as his name suggests, came from Germany. It is usually open and worth a visit for the stained glass and old woodwork.

The church stands prominently upon the hilltop, with the remains of a once-lovely memorial garden running down to the street corner. The hill has a history of its own.

Known as **Agitation Hill**, it was the site where diggers grouped together to air grievances. The Government Camp Reserve was just across Barkers Creek, the Goldfields Commissioner's Office alongside Police Barracks and Courthouse. Miners gathered where they had the visual advantage over the camp, to express their complaints, chiefly over the manner of issuing mining and liquor licences, and the habitual cronyism among officers of the law and some local businessmen. Many of the agitators were dismayed that the Church of England was allotted the site when the town was surveyed in 1853. Local historians see these gatherings, together with those on other goldfields such as the Monster Meeting at Chewton and Red Ribbon Day in Bendigo, as the beginnings of Social Democracy in Australia, leading to the ultimate expression of the Eureka Stockade in Ballarat three years later.

CAMPBELLS CREEK

The Midland Highway from Castlemaine leads under a masonry railway bridge (wondrously complex brickwork to match the angle of the railed to the road) and southward to Campbells Creek. This creek is formed by the confluence of Forest Creek and Barkers Creek, and flows on to join the Loddon River near Guildford.

Campbells Creek township has always been an extension of Castlemaine township and is the site of the main cemetery for the district. Apart from some picturesque buildings along the road, there is a large restorer's barn, and a well-known, sizeable book-barn. There are interesting walking trails through the remains of old mining operations.

Above: Theodore Halacas, Castlemaine Gaol
Opposite top: Katherine Seppings, Lyttleton St
Opposite bottom: Joseph Kinsela

CHEWTON

From the Calder at Elphinstone, the Castlemaine road soon crosses the railway and descends quickly through a winding series of gullies. Before long the outskirts of Chewton are seen, miners' cottages hugging the road, dug-over creek flats to one side or other. Old chapels, shop fronts and schoolhouses appear as the road snakes over and around little hills. This was the Forest Creek Diggings, active from late in 1851, and the world's richest shallow alluvial goldfield.

The town is a typical remnant mining-village, showing the linear pattern of settlement along watercourses where gold was being extracted in the early years of the diggings. Mining was at first an individual exercise, each miner working his small claim – or pooling efforts with his mates to dig and wash more effectively. The pattern changed when large companies entered the field during the 1860s, sluicing areas of hillside and gully and filtering the residue mechanically. As this happened, many traces of workings from the first decade disappeared, and some miners' settlements as well.

From 1874, water was reticulated from Malmsbury to help sluicing operations, which continued until the mid-20th century. Stony and clay soil – where gold was likely to be found - was washed from the surrounding ranges. Red basalt soils were not usually touched by these operations. Today light stringybark and ironbark forests grow on depleted hillsides, with undergrowth composed of hardy native varieties and persistent exotic weeds and herbs. The trees we see today are mostly small and scrubby, being regrowth from the roots of larger trees that cut for firewood, or for props in tunnels and mineshafts.

The remains of all these activities can be discerned from the main road through Chewton. Just behind the houses, pubs and shopfronts lie the actual mining sites. Information about moments of history, buildings in the township and suggested walks around former mines, is available from local Visitor Information Centre.

The centre of **Chewton** is the group of buildings comprising the **Red Hill Hotel**, a pretty redbrick **Post Office** (1879), adjacent **Town Hall** (1861) and the **Primitive Methodist Church** (1861) opposite. The church's twin flying buttresses are quite a spectacle.

The hotel has the best story in town. When it was being built as replacement for an earlier pub, extensive cellars were dug for secure liquor storage and for a local lock-up. Enough gold was found during the excavating to pay entirely for the new building. So the publican added on an assembly hall for local citizens.

Beyond Chewton is **Wesley Hill**, named after the chapel on the left side of the road.

Above: Katherine Seppings, Chewton Butcher and Baker
Opposite: Daryl Chibnall, Chewton

124.

HARCOURT

The northern hills of the Dividing Range were the scene of the fabulous Mount Alexander goldfield with many settlements remaining in the gullies that surround the substantial town of Castlemaine. However the Calder sets its course by the ridges that lead past Mount Alexander itself, sweeping through the orchards and vineyards of Harcourt Valley.

Apart from its natural endowment of good soils and benign climate, this valley has had the advantage of a water supply system for irrigation since the 1870s. It is a highly productive area, notable for apple and pear production, and more recently for cherries and several vineyards.

The valley runs north-south along the crest of the granite ridge beginning at Faraday. On the western side is a tumble of granite slopes covered with red gum and yellow-box trees. On the eastern side the range rises to form the flank of Mount Alexander, the highest eminence in the northern hills of the Dividing Range.

Harcourt has long had a reputation for cool climate fruits and there is plenty of opportunity to sample from roadside stalls and packing sheds in season. The village of Harcourt is at the intersection of the Calder Highway and the Midland Highway coming from Castlemaine. Here also the railway arrives alongside the road on its way to Bendigo.

Near the village is **Apple of the Valley**, a large outlet open seven days a week where the entire range of locally-grown apples and pears may be

purchased in season. In the village is **Harcourt House Antiques**, an imposing limestone house where antique furniture and bric-a-brac is displayed in a domestic setting. At the **Fortis Fruit Shop** on the Calder Highway there is every variety of apple and pear, available in season; also apple jelly and marmalade, quince jelly and other delicious condiments.

Wineries of the area demonstrate the Harcourt cooler climate character. **Harcourt Valley Vineyard** is a long established winery at the southern end of the valley. Nearby, on **Blackjack** Road is the winery of the same name. **Mt Alexander Winery & Cidery** is past the northern end of the village, with distinctly local product that should be sampled. On the southeast flank of Mt Alexander is **Langanook Winery**, reached by following signs to Sutton Grange.

A short distance down the Midland Highway toward Castlemaine is **Skydancers Orchid & Native Plant Nursery**, with an interesting display garden. The nursery is open from mid-week to weekends.

Above: Geoff Hocking
Opposite top: Daryl Chibnall, Harcourt
Opposite bottom: Daryl Chibnall, Harcourt

Gary Chapman, Guildford

MALDON
MT TARRANGOWER

Maldon was built on the slopes of Mount Tarrangower, right where gold nuggets were picked up off the hillside. Tarrangower is a straggly big hill, quite prominent in the last low ranges of central Victoria at the edge of the vast plain spreading north to the Murray River. This section of the range is mainly composed of upturned sandy shales and slate, with dramatic granite domes on the western flank.

Tarrangower rises 571 metres above sea level and is at least half that height above the surrounding countryside. On top there is a lookout tower made from an old poppet head from a Bendigo mine, brought here by train in 1924. The view is wonderful: across plains and ranges from the Dividing Range and the potato-shaped peaks around Daylesford to the endless northern horizon. The tower is illuminated at Easter and can be seen from 50 km away. The mountain is the focus for a Hillclimb event, held annually since 1928 (with a few breaks) – the longest running motor sport event in Australia.

The yield of gold taken from the bowels of this mountain is astonishing. Some of the richest seams in the country were mined in shafts which can be seen just west of the Newstead Rd. Here are the workings of the **North British Mine**, site-works and tunnels and the kilns where quartz was heated until brittle, before crushing to release the gold. The **Grand International Quartz Mining Co** dug Carman's Tunnel which can be seen on guided tour.

The town of Maldon is one of the many towns that grew with the great alluvial mining rushes of the 1850s, and survived beyond that golden decade when attention turned to quartz reef mining. Until the 1870s quartz and alluvial mining employed roughly equal numbers of men. After the mid 1870 quartz reef mining continued to flourish and was profitable until the 1920s. When the North British Mine closed in 1928 it had produced more than 242,000 ounces of gold.

Above: Daryl Chibnall, Maldon
Opposite: Daryl Chibnall, Maldon

THE TOWNSHIP

Maldon is a pretty town that has not yielded totally to the surveyor's pencil. So the main street, called Main St, curves around in wayward fashion, some would say as a mining town's main drag should. Every little street yields an interesting view – of an old chapel on a hillside, or a tall industrial chimney. It is a town for browsing – there are antiques, books, excellent cafés, and plenty of atmosphere.

The old market building, once a white elephant, is now used as the town museum. **Maldon Hospital** is a wondrous building standing on its hilltop with great presence, all pilastered and porticoed.

The **Anglican Church** in High St (1861) is a prime example of the local ragstone with dressings of grey granite. The interior is quite beautiful, with scissors-truss roof and excellent stained glass. Other little churches and chapels dot the hillsides. Also on High St there is a sumptuous mansion built flush onto the pavement, next there is a little building with fantastic patterned brick-nogging around the door and windows.

At the far end of Main St is a tall chimney marking the remains of the **Beehive Mine**. It stands 30 metres high. These extensive industrial remains can be toured.

One highlight, unique in the region, is the steam trains. Maldon Station in Hornsby St is the home of the **Victorian Goldfields Railway**, which runs steam-hauled trains on the line as far as Muckleford North on Wednesdays and Sundays.

At the back of town, beyond the Grand Union Mine, is the picturesque cemetery - with a gate lodge (1866) and sentinel cypresses. It is a place of loving care, as relatives still paint the railings of family tombs. A brick funerary oven stands in the Chinese section for the ritual roasting of food to honour the dead; unfortunately the Chinese grave markers, being timber, were all burnt in the 1969 bushfire.

Above left: Gary Chapman, Maldon
Above right: Gary Chapman, Maldon from Anzac Hill
Opposite top: Katherine Seppings, Maldon
Opposite bottom: Katherine Seppings, Maldon

BENDIGO

After Harcourt Valley the Calder continues to Ravenswood, site of the earliest homestead in the district. The highway continues through hills dotted with red gums and granite boulders, finally climbs over the last big hill (called Big Hill, naturally) into the southern suburbs of the greatest goldfield of all, Bendigo.

The level landscape of **Kangaroo Flat** gives away the fact that this is the edge of the great plain of the Australian interior. This is where the endless horizontal inland of the continent begins. The main road passes through typical modern suburbs that were in fact early mining sites. Today some remnant older buildings survive to tell the story: little redbrick churches and schools, former stables and coach-builders' shops.

Soon the most prominent structure in Bendigo, **Sacred Heart Cathedral**, comes into view on the hillside above the main road. First the 87-metre spire, then the impressive façade loom over the neighbourhood before the road veers northeast toward the city centre. On the right is the poppet-head of **Central Deborah Mine** and a tramline enters the street from the adjacent tram terminus.

At the turn into High St, the view is toward the **Alexandra Fountain** and the towers and domes of **Pall Mall**, the heart of Bendigo. The **City Family Hotel** has a long frontage to the street, facing former insurance buildings and banks on the left. A glimpse up the steep rise of **View St** reveals an impressive row of classical buildings.

The tramline skirts the fountain and takes the centre of the grand avenue ahead.

The central view of the city used to be dominated by mining operations, especially by Hustler's Royal Reserve Gold Mine just beyond the fountain. The poppet-head and crushing plant were replaced in 1921 by the **Returned Soldiers League Hall**, with its frontage a domed belvedere intended for use as a bandstand.

Deep-lead mining continued below the actual city centre until 1954 - and is soon to resume in the belief that as much gold remains as has already been taken from the ground. In the towers and high roofs of the grand buildings can be seen the signs of prosperity in the Victorian Era. The golden years of the reign of Queen Victoria coincided with the golden years of the Colony of Victoria, named after the queen at the separation from New South Wales in 1851. A few months after its foundation the new colony had the extraordinary good luck to pick up enough gold off the ground to give it strong economic foundations.

Above: Daryl Chibnall, Kangaroo Flat
Opposite: Daryl Chibnall, Bendigo

134.

Bendigo stands on one of the richest goldfields in history and it's just down the road from a dozen other goldfields. Today Bendigo combines the energy and optimism of one of the nation's fastest-growing urban centres with the gracious framework of a handsome and well-planned city of the 19th century.

Within the shadow of its pinnacles and crested roofs is a modern business centre, but Bendigonians are very aware of the special character of their town, and fiercely protective of their particular historical inheritance. No other town in Australia has spent so much time and care on rendering its past accessible and enjoyable.

There is much to see and experience, walking around the streets and seeing close-up the splendid buildings of a past era, when an Australian country town sought to catch up with Europe in four decades. The grand buildings and their magnificent interiors demonstrate the high seriousness of those who designed and built them, as well as their eye for beauty.

The buildings still carry a sense of pulsating life, of a time one hundred years ago when everyone was an adventurer, a failed digger, or an entrepreneur scooping up a fortune from those who had their noses to the ground. There were pubs, music halls, art galleries, stock exchanges, poor-houses and orphanages, Chinese cemeteries – all built from the wealth wrested from holes in the ground. From a world that was bright and hot outside, dark, wet and dangerous down beneath.

To explore Bendigo:
- take a tram ride down Pall Mall
- get in the elevator at Central Deborah and go down a mineshaft into the dark unknown
- rest in the Chinese Gardens, the wander through the museum next door and look Sun Loong - the largest Imperial Dragon in the world – in the eye
- be dazzled by the golden ornament inside the Old Town Hall
- explore the treasures of the art gallery

CENTRAL DEBORAH GOLDMINE

The Central Deborah Goldmine just off High St has operated since the 1850s, and was the last mine to close in 1954. The poppet-head can be seen above surrounding buildings to the south of High St. Bendigo Council purchased the mine with all its functioning machinery in 1971 for $6000; it is now run as a tourist experience, taking visitors twenty storeys below ground level to see shaft and tunnel mining at close quarters.

Above: Christine Ramsay
Opposite top: Daryl Chibnall, Bendigo
Opposite top: Katherine Seppings, Central Deborah Goldmine

136.

PALL MALL TRAMS

Bendigo's current tram system is part of a larger system dating from 1890 – steam at first, electrified in 1903 – that went from Golden Square to Emu Point, and from Eaglehawk to Quarry Hill, crossing at Charing Cross. Today it travels from **Central Deborah Terminus** to **Lake Weeroona**, a pleasant return journey of less than an hour past some of the city's finest sights. Run as a tourist venture by the City Council since 1972, when the system was threatened with closure, the Bendigo tramway is now the home for early tramcars from Melbourne, Adelaide and Sydney. It has some of the oldest and rarest vehicles surviving in the world.

Early bogey-cars and four-wheelers make up the stock - many of them are in running order and others are on view in the Tram Museum. What are called trams in Australia (and the rest of the former British Empire) were originally horse-drawn carriages set on rails. In many places these evolved into steam-powered operations, for example, in Sydney, Hobart and in Bendigo. Then, by 1900, electric transport was all the go.

The advent of electricity brought about self-propelled vehicles, an American invention known there as streetcars or trolley-cars. A cheap form of railway running along the streets, they soon spread to any city of a certain size, but in Australia they came to be a necessary part of a state capital or a major provincial town. In Victoria, tramway systems operated in Geelong, Ballarat and Bendigo from the earliest years of the 20th century. Melbourne itself had horse-tramways from at least the 1870s, followed by cable trams (as in San Francisco) for many years until electrically powered vehicles took over all the routes. Apart from Melbourne, Bendigo's are the last trams in this country to be seen running along the streets.

Above top: Katherine Seppings
Above bottom: Katherine Seppings
Opposite left: Christine Ramsay, Pall Mall
Opposite right: Christine Ramsay, Pall Mall

138.

VIENNA-IN-THE-BUSH

Bendigo is still the finest goldrush city in Australia - with just an edge over Ballarat, Victoria's other great goldrush centre, and Kalgoorlie in Western Australia. Victoria's gold rush cities are probably the best remaining from the 19th century in the world. While Bendigo is an Australian country town with a population only now approaching 100,000 people, the city centre has strong echoes of European cities like Paris and Vienna.

Outside the centre, the rest of the city is resolutely Australian with boom-style mansions on the central hills, rows of brick houses with shady verandas, and sprinklings of timber miners' cottages in the oddest places, marvellous big churches, and charming pubs like **The Goldfields** on Marong Road. There are few completely fine streets; all is chaotic, the legacy of mining boom-and-bust and its attraction for both rich and poor. And witness to the fact that, until at least the 1950s, mines and scarred hillsides were to be found in many odd places throughout the city.

However, there is no denying the panorama of fine buildings at the heart of the city. The profile is of domes, towers, crested Mansard-roofs, classical pillars and motifs, Gothic spires and pinnacles. There is a surprising voluptuousness redolent of the fabulous wealth that came from beneath the very foundations of these buildings. Bendigo's opulent physical character is the direct outcome of a gold boom that began in 1852 and did not fade away for the whole of the 19th century.

The present-day needs of Bendigo as a major regional rural centre have impinged on the historic city of the goldrush. Most of the commercial heart of the city towards the railway station has been completely renewed in the last three decades – in a resolutely uninspired way. New commercial buildings have leaked out of Mitchell St and now confront the elegant 19th century atmosphere of Charing Cross with its magnificent fountain.

Sophistication came early to Bendigo. It arrived with the German-born architect William Charles Vahland in 1857. Vahland's buildings generally speak a classical architectural language with a baroque or mannerist accent. His are the defining buildings of the town centre – the Town Hall, Capital Theatre, Institute of Mines (CAE), City Family Hotel, Alexandra Fountain, hotels and shops.

Above top: Katherine Seppings, Town Hall
Above bottom: Katherine Seppings, Town Hall
Opposite top: Katherine Seppings, Sacred Heart Cathedral
Opposite bottom: Katherine Seppings, Pall Mall

VAHLAND & GETZSCHMANN

The architectural firm Vahland & Getzschmann designed so many of the dwellings, schools, churches, pubs and commercial buildings of the city that this one firm can be said to have stamped the essential character of Bendigo.

Of course, many other architects have made contributions to the city. Quite a few of the profession in Melbourne had a chance to design a factory, bank or insurance house. Names occur such as Charles Webb, Smith & Johnson, Leonard Terry and Joseph Reed - Melbourne's best. Reed's partner WB Tappin designed Sacred Heart Cathedral in the 1890s. The (mostly anonymous) Government Architects built administrative premises – a succession of post offices and courthouses, the gaol, public schools. WG Watson was Chief Architect in the Office of the Public Works and had the good fortune to design Bendigo's final glorious Post Office and Courthouse during the 1880s.

The work of several local architects remains to preserve their names. RA Love designed the first Council Chambers (now buried inside the 1885 building) in an eccentric brick manner. His Cathedral of St Paul in Myers St is a spirited redbrick Gothic construction with its own eccentric characteristics. John Beebe built a number of brick and stucco buildings late in the century in a distinctive Arts & Crafts idiom, such as the YMCA (High St), and his brother William designed the Old Fire Station (View St).

What Bendigo still has, which few other towns can boast, are some splendid interiors to match the grand exteriors of several city buildings, in particular:
- **Old Town Hall**, between Hargreaves and Lyttleton Sts
- **Supreme Courthouse**, Pall Mall
- **Shamrock Hotel**, Pall Mall
- **Mully's Café Gallery** (former National Bank), Pall Mall
- **Bendigo Art Gallery**, View St
- **Bank-Bar on View** (former Union Bank), View St
- **Capital Theatre** (former Masonic Lodge), View St

- **Chinese Joss House** (temple), Emu Point – end of tramline
- **Golden Dragon Museum, Gardens & Kuan Yin Temple**, Pall Mall & Bridge St

Local churches are also worth seeing for their embodiment of Bendigo's spiritual habits and aspirations. The fine group of churches in the Forest St, View Hill area is second to none. Worthy of mention apart from this list are the large Methodist churches (now Uniting) found at Golden Square and Long Gully; these reflect the link between Methodism and the self-made businessman and trade-unionist of the 19th century. St Paul's Cathedral (Anglican) and Sacred Heart Cathedral (Roman Catholic) are a study in contrasts, one richly furnished, the other austere and spacious.

Above top: Katherine Seppings
Above bottom: Katherine Seppings
Opposite top: Joseph Kinsela, Town Hall
Opposite bottom: Katherine Seppings, National Bank / Mully's Café

142.

BUILDINGS & PRECINCTS

POST OFFICE & LAW COURTS

Bendigo's centrepiece is the pair of civic buildings standing prominently on Pall Mall's broad avenue. Built of brick, with cement-stucco facing and ornamentation, in a French Renaissance style, double-storied with high Mansard roofs and truncated domes at the centre they appear identical. But the Post Office is slightly larger and has a clock tower, while the Courthouse has an extra storey at the centre above its vestibule and the Supreme Courtroom.

Like the Town Hall, which they emulate in size and general appearance, each building stands entire, monumental, to be viewed from all four sides. Raised on a bluestone podium, they each have a deep basement level. The Post Office's was for secure storage and handling of precious cargoes, the Courthouse's for the confinement of persons being brought to justice. Solid basalt and cast-iron fences surround the buildings, securing an access of light and ventilation to the basements; these fences are punctuated by bluestone corner-piers with elaborate light standards supported by chimeras.

The **Post Office** is no longer used by Australia Post, but serves as a rather splendid **Visitor Information Centre**. Museum exhibits on historic themes occupy the former postal hall and offices. The town clock chimes quarters and hours, to a curious sequence known only to Bendigo; the bells were cast by a foundry in Castlemaine.

Entry to the **Law Courts** is through the arched loggia where notices are displayed. A side entrance leads to a small vestibule from where a glimpse is gained of the first of two superb double flights of stairs ascending to the upper vestibule. Tall arcades rise on three sides of the stairway giving onto the waiting areas of the vestibule. All is finely detailed with Classical ornament of the Corinthian Order. This is one of the grandest public spaces in Australia. The Supreme Court room is of a similar scale, with matching detail and two immense bronze gasoliers.

OLD TOWN HALL

There was already a municipal chambers from 1859 on site when, in 1885, WC Vahland began his greatest work: the main hall and all four facades of the Town Hall. Completely enclosing the earlier structure, the building has a two-storied classical scheme. There are paired Tuscan columns on the lower storey, Corinthian above, and open pediments alternately angled or curved. Everywhere, there are large arched windows.

Each corner rises into a truncated pyramid with attic windows, except the north-west corner where there is a substantial tower 36.5 metres high. Muscled Atlas figures crouch to take the weight of the clock face – but no clock is there. It is said to have been left out because the Post Office clock tower was being built within the same year.

Nothing can prepare the visitor for the impact of the newly-restored hall within the building. There is a vast amount of gold, and original murals of mythical female figures and cherubs and other decorations by German artist, Otto Waschatz. There are gilt plaster cherubs above each doorway, a coffered ceiling with gilt pendants, and Gothic arches above each window. It is overwhelmingly lovely, and expresses in an instant the fabulousness of the gold boom. Tours are a must; make inquiries at the Visitor Information Centre.

Above: Katherine Seppings, Town Hall
Opposite top: Katherine Seppings, Post Office & Shamrock Hotel
Opposite bottom: Katherine Seppings, Town Hall

144.

SCHOOL OF MINES

The tower of the School of Mines is the third prominent landmark of the city centre after the Post Office and Town Hall towers. The impressive group of cement-stucco buildings reads as two distinct parts, the main entrance in a two-storied façade of 1879 with a central arcaded loggia, and the three-storied block with tower of a decade later. The earlier section is soberly classical, while the later is an example of Vahland's German 19th century manner – where refined classical elements are reduced to a thin surface enriched by elaborate friezes.

This fine institution was part of a social movement finding a foothold in every new community in the country during the mid-19th century. Mechanic's Institutes and Schools of Mines were an expression of the idealistic move to offer the working-man facilities for self-improvement. An aspect of the growth of Social Democracy in mining areas, this movement, together with the principles of trade unionism and universal franchise, was an opening door for each person to have some control over his or her own destiny.

In Bendigo, the first facility erected at the Mechanic's Institute was a timber exhibition and dance hall, later expanded by lecture and meeting rooms. Pressure for education in the techniques of mining led to the establishment of the School of Mines in 1871; this was expanded over the next few years and merged with the Institute.

In 1887 an unusual Octagonal Library was added behind the entrance front. For its time a capacious facility, the library has a centralised form with gallery following the classical orders – base level Tuscan Doric, first floor Corinthian – with enriched cornices and a shallow dome. This is an elegantly beautiful interior retaining its original gaslight-fitting, now electrified. It is now used as an occasional restaurant, a training facility for students of catering and hospitality courses. It is worth finding a time to visit one of Vahland's most surprising interiors.

SHAMROCK HOTEL

The Shamrock Hotel is deservedly famous as one of the grandest hotels in Victoria. Standing opposite the Post Office tower, it complements the style of the civic buildings with its French Second Empire Mansard roofs and grey cement-stucco surface treatment. There is an international flavour in its High-Victorian massing and assertiveness, although it was designed in Bendigo in 1897 by Philip Kennedy, a pupil of W C Vahland.

Standing four storeys high plus attics, the Shamrock has an ornate veranda for the two lower levels and overall the bravado face of a European or American resort hotel of the late 19th century. The entrance is through Italianate arches on curly cast-iron columns, where the magnificently tiled floor leads the eye to an ornate staircase. Interiors show a similar splendour and have been wonderfully renovated for today's needs. It is built in high style, no expense spared, and yet it was constructed during the course of the worst depression the colony had ever experienced.

The Shamrock Hotel has played an important role in Bendigo's social and cultural life. The first building on the site was a restaurant and music hall started in 1852. When it sold two years later to two Irishmen, the name changed from Theatre Royal to The Shamrock. From 1859 it was the office for the Cobb & Co stage coach and telegraph company. A much larger hotel was built the following year, which during its 35 years received many famous visitors to the city, including on 23 November 1861 the exhausted explorer King, only survivor of the Burke & Wills Expedition. This second Shamrock was entirely demolished to make way for its 1897 replacement.

Above: Katherine Seppings, School of Mines
Opposite top: Katherine Seppings
Opposite bottom: Katherine Seppings, Shamrock Hotel

COLONIAL BANK

Erected in 1887 for the Colonial Bank and later serving as the National Bank, this small yet palatial building was designed by Vahland on the model of a double-height Roman triumphal arch - or one of the stone tombs of Petra. There are ground level rusticated Ionic columns, first storey Corinthian columns, an attic storey with female Terms supporting exaggerated cornices with huge urns and a broken pediment. Altogether, it is one of the most sensational facades in Australia.

The interior does not disappoint – with a high (9.5 metres) coffered ceiling with light panels at the entre, and elaborate pilasters on the walls. There is a stairway to a balcony overlooking Pall Mall. Today it is **Mully's Café**, a busy eatery with pictures for sale. Pleasant though it is, the business looks out-of-place in such a sumptuous chamber.

BEEHIVE BUILDING & STOCK EXCHANGE

A handsome three-storied building in Pall Mall facing the RSL Hall, the Beehive Building's restrained façade conceals the triple level-arcade behind. Built as the **Bendigo Stock Exchange** after a fire in 1871, it was designed in Melbourne by Charles Webb, architect for the Windsor Hotel and the Royal Arcade.

The **Beehive Stores** operated from 1854 to 1987 at street level, giving the building its name. The Bendigo Stock Exchange was for a time the busiest in the Australian colonies, where multi-millions – *even billions* – of dollars worth (in today's currency) of business in goldmining stocks changed hands. The building is currently being offered for redevelopment, as a scheme which retains the galleried interior and decorative elements.

PALL MALL

This street is just about grand enough for its name, and it is a delightful place for a stroll: to enjoy fine buildings, public statuary, trams, trees, not to mention food and drink at the pavement cafés.

Some of Bendigo's most important commercial businesses have a Pall Mall frontage. The Myer Emporium is an unabashedly 1960s façade, rebuilt after a fire and extended into neighbouring stores, but this is a very important part of the history of Myer. In the late 1890s Simcha Baevsky arrived in Bendigo from Russia; he hawked merchandise in heavy suitcases until he could open his first drapery store in View St. Other stores soon followed in Hargreaves St, Pall Mall, then Melbourne. And he changed his name to Sidney Myer.

It took nearly thirty years for Pall Mall to assume the present form of a broad central avenue for the city. Like many goldrush towns, the centre of Bendigo was a chaotic heap of mud – literally diggings – in the earliest years. Richard William Larritt was given the task of surveying the township of Sandhurst, to straighten wagon-tracks and move huts and tents aside so roadways could be clearly defined. This done, allotments could be offered for sale and a permanent settlement encouraged.

Larritt's vision was for an avenue with civic buildings backing onto Bendigo Creek. But pressure from council and public for housing lots on the area delayed his plans. He played a waiting game, and a severe downpour and flood proved its unsuitability for housing. The creek was given a wide masonry channel and the banks were built up. Larritt's vision was finally realised in 1882 when the government began planning the present Post Office and Courthouse buildings.

Above: Katherine Seppings
Opposite: Chris Kirwan, Stock Exchange

148.

PRINCESS ALEXANDRA FOUNTAIN

The Princess Alexandra Fountain forms a striking element at Charing Cross, where High St intersects with View and Mitchell Sts and Pall Mall begins. Designed by Vahland and built of polished Harcourt granite. The great bowl is a single piece and the total weighs 20 tonnes. Local contractors provided stonework, lighting and plumbing, while the figures were modelled in marble and bronze by E Temper of Melbourne. The fountain was declared open in 1881 by the Princes Albert and George, sons of the Princess and Prince of Wales (later Edward VII).

CITY FAMILY HOTEL

The handsome City Family Hotel was designed by Vahland & Getzschmann for Jean Baptiste Loridan, a local miller and investor. It has been one of Bendigo's leading establishments and for many years housed the collection of artefacts, banners and other items of the Chinese community, now in the Chinese Museum.

In the late 1960s the ornate double-storey veranda was removed, along with the balustrade and urns of the distinctive tower. Signage now disfigures the tower and Pub Tab decor has infested the former lavish Victorian interior. The superb triple flight corkscrew staircase survives as witness to the original quality of this landmark building.

The neighbouring headquarters of Bendigo Bank, an admirable and venerable institution of this city, demonstrates the inability of much recent commercial design to compare with the best work of the 19th century. Perhaps in time, the Bank will take opportunity to reconsider the frontage it presents to its home city.

VIEW ST

View St rises steeply to the north of Alexandra Fountain. Superb lighting poles at the centre bring back the early character of the street. Lined with former banks and insurance houses, it was for 100 years the financial heart of the city, close to the Stock Exchange. On the right is the National Bank, the only survivor in its original premises. The former post office is now Sandhurst Trustees; then in order up the hill - Temperance Hall, Penfolds Gallery (former hotel), and Trades Hall.

The **Bendigo Art Gallery** stands newly revealed, with its entrance and extension in the shade of the magnificent Capital Theatre. Opposite is the Natural Resources & Environment Building, occupying the former site of Bendigo's Princess Theatre. Demolished in 1970 for a petrol station, the Princess was in its heyday one of the best theatres in the colony. Among other greats, Dame Nellie Melba performed here several times between 1886 and 1913. Remodelled in 1938 in Art Deco style as a cinema, the theatre survived until the advent of television. Its demise is perhaps Bendigo's greatest single loss to progress.

A fine array of frontages, some with ornate verandas, climbs the hill on the left side of View St. At the top of the rise is the **Rifle Brigade Hotel**, a handsome pub with a good eatery and boutique beers.

Above: Katherine Seppings, City Family Hotel
Opposite top: Katherine Seppings, Princess Alexandra Fountain
Opposite bottom: Katherine Seppings, View St

150.

UNION BANK

The former Union Bank on View St is one of the most striking Classical fronts in Bendigo consisting of a Corinthian portico-in-antis that is expressed as an internal space between the side walls instead of an external covered area. Smith & Johnson architects (from Melbourne) designed the bank in 1875 to replace the earlier building of 1857. The small vestibule leads to a magnificent chamber roofed internally as a square dome, with deep coving breaking at three handsome windows high in the rear wall. There is little ornament except for a hugely glorious rose at the centre of the ceiling and a fireplace at one side. The effect is in the proportions.

The present owner plans after restoration to open the bank as a wine bar. At the rear is a two-storied accommodation block for single members of bank staff, possibly as a deterrent against robbery; these are now available as serviced rooms. Beyond is the original smelter house with chimney; now an office. Banks on the goldfield used to smelt gold into ingots as a secure means of preparing it for transportation to Melbourne; the miner's weight of gold dust or nuggets was first recorded in an inventory. This property retains all the parts of a working bank on the goldfield, and is a rare survival.

BENDIGO ART GALLERY

The Bendigo Art Gallery was opened in 1890 in the former premises of the Bendigo Rifle Brigade, whose interior was refashioned into a series of handsome galleries. An unlovely yellow brick entrance was added in the 1960s; its removal in 1998 along with the neighbouring ANA Motel allowed space for a new entrance/administration area alongside the Capital Arts Centre and restored the aspect of the 140-year-old polychrome-brick frontage.

CAPITAL THEATRE

Built in 1875 as the Masonic Hall to the design of Vahland & Getzschmann, the Capital Theatre was saved from threatened demolition in 1990 and gradually refashioned into Bendigo's Performing Arts Centre.

There is a Baroque ripeness in its classical manner. The Corinthian portico - the most imposing in Victoria - is set halfway into the façade of the building. It is all in stuccoed brick, resembling stone. From the top of the rise, the Capital dominates View St. Inside, the elegant foyer leads to a stair hall, left of which is the Banquet Room and the very beautiful (though not large) Bendigo Bank Theatre, formerly the Masonic Room. Recent redecoration has revealed the classical décor —

Vahland, a member of the Freemasons Lodge, shows how fully he understood the ancient art of building.

Upstairs the Theatre was originally a ballroom and occupies the full frontage of the building. Vahland was asked in 1890 to add a proscenium and make a stage using the building next door. The auditorium has a German-Bavarian character of the 18th century. Walls are blank-arched in a baroque manner, with pilasters that break into heavy capitals representing Beauty and the Beast. Seating is for almost 500 people.

This theatre is the principal performance venue for Bendigo, with a year-round programme of visiting professional companies and the best of local drama and music. The Capital Theatre is run in conjunction with the neighbouring **Old Fire Station** and **Dudley House** (further up the hill) as a variable group of meeting rooms and facilities, ideal for parties and conferences.

Above: Katherine Seppings, Fire Station and Gallery
Opposite top: Katherine Seppings
Opposite bottom: Katherine Seppings

ST PAUL'S ANGLICAN CATHEDRAL

A large redbrick church in an assertive Gothic style, St Paul's was built as a parish church to the design of goldfields architect RA Love. Consisting at first of a short nave with lofty side windows, the church was extended in 1890 by large transepts and sanctuary, making it Bendigo's largest Anglican church. A tall tower with strong corner pinnacles marks the position of the cathedral at the eastern side of the city centre. The peal of eight bells is regularly rung.

The interior is distinguished by a spiky arch-braced timber roof and magnificent furniture; the stained glass is especially bright and cheerful. St Paul's has long been known for its musical tradition and magnificent organ. The church was designated cathedral for the Diocese of Bendigo in 1981; the adjacent hall has been renovated as offices for the Diocese.

SACRED HEART CATHEDRAL

Sacred Heart Cathedral is one of the largest cathedrals in Australia, dominating the entrance to the city centre from Melbourne. Designed by Victorian-born architect William Brittain Tappin for the Roman Catholic Diocese of Sandhurst, it took over 80 years to complete. The form and details of the cathedral are identifiably Early English Gothic. The entrance and nave, opened in 1901, were in use for more than half a century before increased income from real estate during the 1960s and 70s allowed the completion of the cathedral. Transepts, presbytery and chapels were built according to original plans over twelve years from1966, doubling its size. The summit cross of the huge (87 metre) central spire was put in place in 1977, making it the third highest in Australia. With such size and eminence, Sacred Heart Cathedral is used for many important occasions. The austere, light-filled interior lends itself to musical and dramatic presentations as well as to grand religious ceremonies. Of especial interest is the west window, from Hardmans of Birmingham (UK); a large organ stands to the sides of the west gallery. An immense painting of Christ and Stations of the Cross are rather dominant in the nave aisles – a particular Parisian religious style of the late 19th century. This very spacious interior will take time to acquire great works of art and furnishings commensurate with such a building.

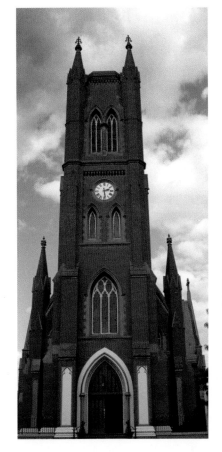
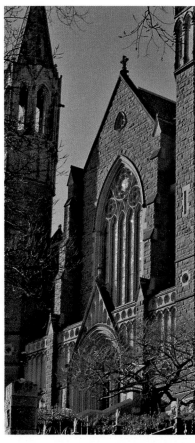

Above left: Katherine Seppings, St Paul's Anglican Cathedral
Above right: Katherine Seppings, Sacred Heart Cathedral
Opposite: Theodore Halacas, Sacred Heart Cathedral

154.

CHINESE JOSS HOUSE, CHINESE MUSEUM & CENTRE

For the Chinese it has been a long journey from discriminated minority to having a respected profile within Bendigo. There were more than four thousand Chinese miners at the height of the rush, with many coming to make a fortune and returning to their families in China. The clear role and popularity of the **Golden Dragon Museum** in the city's cultural presentation signals the modern identity of the Chinese community. The museum and associated **Chinese Gardens** and **Kuan Yin Temple** have been created since 1991 on the site of old Chinatown in Bridge St, with a ceremonial entrance gate from Pall Mall.

The ceremonial dragons have long been a part of street processions in the city, especially the Easter Festival (dating from 1871). These dragons are spectacularly displayed in the circular museum, **Sun Loong** taking several revolutions of the building to contain his great length (over 300 metres). Bendigo and the Easter Festival have long been a focus for Chinese families throughout Victoria, bringing large numbers who participate in the street processions with the two Imperial Dragons and attendant figures. The splendour and intricacy of these handsome figures and masks can be seen up close in the museum.

The **Joss House at Emu Point** (at the end of tramline) is a survivor from early times, being built in the 1860s. It is in three sections - the temple, a

caretaker's residence and an ancestral hall. The god worshipped here is Kuang Gung, god of war and one who brings peace and prosperity. Red and gold are the principal colours of the temple's interior, being the colours of happiness. This building has recently been restored by the Central Victorian Branch of the National Trust, advised by a Chinese historian through the auspices of the Bendigo Chinese Association.

Above: Katherine Seppings, Chinese Pagoda
Opposite top: Katherine Seppings
Opposite bottom: Katherine Seppings

WOODEND TO TRENTHAM

From Woodend the road to Trentham and Daylesford dips into the pleasant valley of the Campaspe River. By the river is **Woodend Berry Farm**, where visitors may pick the summer crops and wander through the maze and lavender garden. Then the countryside rises gradually while keeping in view forested ranges of the Great Divide to the south. Slopes of cleared pastureland with old manna gums and hedgerows of cypress or hawthorn follow the fertile volcanic soils, while stringybark forest remains on stony areas. The countryside could almost be English – the valleys are shallow with clear gentle streams - these are younger land surfaces than are usual in Australia.

Just before the village of Tylden the road crosses Little Coliban River, which originates in the hills of east Trentham. After this, the country rises gently onto a higher tableland. Soils are rich red volcanic loams, good for growing potatoes and other root crops, and for dairy pastures. Trees along the road – ribbon gums and stringybarks - are noticeably taller although some still show scars from the 1983 holocaust.

TRENTHAM

Trentham is Blue Heelers country; adherents of the long-running TV drama series will recognise rolling hills with groups of tall manna gums. **Trentham Falls** and **Gorge** are worth visiting to see the thirty-metre tall blackwood trees thirty metres tall; here the Coliban River falls over the end of a six million year old lava flow onto a rocky base dating from 270 million years earlier. The falls were once harnessed for power by a turbine, that drove a sawmill.

Also recognisable is the **Catholic Church of St Mary Magdalen**. On a shrubby rise at the entry to the town, this redbrick building is used as a setting for funerals or weddings in the TV series. Although Gothic with high pointy roof, it has a resolutely flat-roofed porch and annex. However, to enter the church is to be surprised by the lovely interior, turned around the 'wrong way'.

High St has picturesque premises, especially **The Fir Tree B&B** (a former guest house), the **Old Bakery**, and a row of timber-fronted shops. The **Cosmopolitan Hotel** has a gem of an 1865 front bar, hanging lanterns and original scrolled wallpapers. Several fine cafés and restaurants, and the excellent **Reverie Bookshop** are also in High St.

Trentham has been an important logging and milling town for most of its life. The closing of the Woodend to Daylesford railway in 1978 and recent reduction of timber-getting in the surrounding forests have left the town a quieter place. However, this very fact has brought new residents and a growing lifestyle change.

Trentham is becoming widely known as a place to browse books, take good meals and poke around the old railway station – now the **Visitor Information Centre**. The town is quiet early in the week, but can be busy at the weekend.

Above: Sandy Scheltema, Powerful Owl
Opposite: Alison Pouliot

DAYLESFORD

Daylesford occupies three sides of a very large hill, so that every street is heading either up or down it. But the views... The town is 600 metres above sea level, high above most of the neighbouring countryside, especially to north and west. Daylesford is a beautiful town, and over the last decade has turned itself into a busy rural retreat with a large number of weekenders and day-trippers. They come to enjoy an extraordinary variety of walks, views, restaurants, cafes, museums, galleries, bookshops, lakes, spas and mineral springs.

DAYLESFORD BOTANIC GARDENS

The town wraps itself around the northern and western slopes of Wombat Hill, at the crest of which is a dramatic profile of mature conifers. These firs and redwoods from Europe and North America are the chief glories of the **Botanic Gardens**, dating from 1861. There is also a great Brazilian Monkey-Puzzle in company with later plantings of Australasian rainforest species. The deep red loam of the ancient volcanic cap was deliberately chosen as the site for the gardens early in the town's history.

The selection of plant species was assisted by the famous naturalist and inaugural director of Melbourne's Botanic Gardens, Baron Ferdinand von Mueller. The Daylesford Gardens are one of the most beautiful of their kind, not least for the early planting but also for the Fern Gully laid out by Taylor & Sangster in 1887. As well as giving energetic growth to a wide variety of conifers and flowering shrubs, Wombat Hill also offers the most exciting views in town from the historic all-concrete observation tower (1937). There is a kiosk which serves teas.

WOMBAT HILL

High on the hillside is a crop of spires from the town's churches, the old Primary School of 1874, and the towers on the Old Convent Art Gallery and post office. Next to the old primary school in Vincent St is the **Historical Society's Museum & Information Centre**. On all sides there are charming timber cottages with elegant verandas and well-kept gardens, as well as occasional grand mansions such as **Mount Stuart House** and **The Manse** (both now B&Bs).

The **Uniting Church** (former Methodist, 1867) has the tallest spire, and a good polychrome brick nave and tower. Next to the present Uniting and Anglican churches are the original church buildings, relics of the earliest years of the goldrush. Nearer the hilltop is the former **Presbyterian Church**, a neat redbrick building with an excellent octagonal spire.

Above: Daryl Chibnall
Opposite: Daryl Chibnall

■	Cleveland Winery	Winery	55 Shannons Road	5429 9000
■	Farrawell Wines	Winery	60 Whalans Track	5429 2020
■	Glen Erin Vineyard Retreat	Winery	Rochford Road	5429 1041
■	Moonrise Wines	Winery	206 Grahams Rd	5429 1964
■	Portree Vineyard	Winery	72 Powells Track	5429 1422
■	Ashworths Hill	Winery	Ashworths Road	5429 1689
Langley				
■	Langley Farm Cottage	Accommodation	391 Carstairs Rd	5423 4161
Lauriston				
■	Mudlark	Accommodation	9 Darcy Lane	0427 239 170
	Lauriston Lavender Farm	Nursery	478 Lauriston Road	5423 9151
Leichardt				
■	Connor Park Winery	Winery	59 Connor Rd	5437 5234
Little Hampton				
	Caring for You	Health Spa	Lot 6/100 Fern Road	5424 1756
Lockwood				
■	Wooragee Bed & Breakfast	Accommodation	73 Boyds Rd	5435 3470
■	Potiche - Bgo Pottery Complex	Art	44 Gap Road	54353609
Lyonville				
■	Radio Springs	Accommodation	High St	53485562
Macedon				
■	Black Forest Motel	Accommodation	426 Black Forest Drive	5426 1600
■	Ellandee Cottage	Accommodation	53 Ellandee Crescent	5426 2165
■	Lawson Lodge	Accommodation	227 Lawson Road	5426 1551
■	Macedon Caravan Park	Accommodation	324 Black Forest Drive	5426 1528
■	Foodies Roadhouse	Eateries	313 Blackforest Drive	5426 1842
■	Macedon Family Hotel	Eateries	33 Smith Street	5426 1231
	Macedon Nursery	Nursery	Nursery Road	5426 2558
	Mountside Nursery	Nursery	70 Honour Avenue	5426 1480
Maiden Gully				
■	Byronsvale Vineyard	Accommodation	50 Andrews Rd	5447 2790
■	The Stonehouse of Maiden Gully	Accommodation	29 Monsants Road	5449 6602
■	Pratty's Patch	Eateries	35 Monsants Rd	5449 6341
■	Balgownie Estate	Winery	Hermitage Road	5449 6222
Maldon				
■	Agatha Panther's Cottage	Accommodation	87 Church Street	5475 1066
■	Almond Tree Cottage	Accommodation	Church Street	9836 2380
■	Benaird Cottages-Pond	Accommodation	3 Pond Drive	5475 1202
■	Benaird Cottages-Preece	Accommodation	Lot 2 Johns Road	5475 1202
■	Calder House	Accommodation	44 High St	5475 2912
■	Central Service Centre	Accommodation	1 Main Street	5475 2216
■	Clare House	Accommodation	99 High Street	5475 2152
■	Derby Hill Accommodation Centre	Accommodation	Phoenix St	5475 2033
■	Eaglehawk Gully Apartment	Accommodation	35a Reef Street	5475 1449
■	Fairbank House	Accommodation	9 Ireland St	5475 1094
■	Farrington's Cottage	Accommodation	40 Main St	5475 2727
■	Fuller's Cottage	Accommodation	29 Lowther Street	5475 2744
■	Hales Cottage	Accommodation	18 Church St	5475 1094
■	Hardys B&B	Accommodation	113 High St	5475 1027
■	Heatherlie	Accommodation	72 High Street	0413 123 650
■	Heritage Cottages of Maldon	Accommodation	41 High Street	5475 1094
■	Herrmann's Cottage - Indigo Estate	Accommodation	19 Polsue Street	0416 108 222
■	Maldon Caravan & Camping Park	Accommodation	Hospital St	5475 2344
■	Maldon Holiday Cottages	Accommodation	28 Sells Lane	5475 2927
■	Maldons Bluegum Cottage	Accommodation	14 Church St	5475 1094
■	Maldon's Eaglehawk Motel	Accommodation	35 Reef St	5475 2750
■	Minilya Bed & Breakfast	Accommodation	39 Adair Street	5475 1356
■	Miss Cinnamon's Country Cottage	Accommodation	7 Spring Street	9725 4161
■	Mount Hawke of Maldon	Accommodation	24 Adair Street	5475 1192
■	Nuggetty Cottage	Accommodation	30 Nuggetty Rd	5475 2472
■	Palm House B&B	Accommodation	2 High St	5475 2532
■	Peppercorn Cottage	Accommodation	5 Phoenix St	5475 1778
■	Porcupine Township	Accommodation	Cnr Bendigo & Allens Rds	5475 1000
■	Sandgarie Farm B&B	Accommodation	836 -Newstead Road	5475 1528

Terraces Miners Cottage	Accommodation	11 Fountain Street	9818 8512
The Loft and Barn	Accommodation	64 Main Street	5475 2015
Derby Hill Accommodation Centre	Accommodation	Phoenix St	5475 2033
Farrington's Cottage	Accommodation	40 Main St	5475 2727
Heritage Cottages of Maldon	Accommodation	41 High Street	5475 1094
Maldon Caravan & Camping Park	Accommodation	Hospital St	5475 2344
Maldon Holiday Cottages	Accommodation	28 Sells Lane	5475 2927
Maldon's Eaglehawk Motel	Accommodation	35 Reef St	5475 2750
Miss Cinnamon's Country Cottage	Accommodation	7 Spring Street	9725 4161
Nuggetty Cottage	Accommodation	30 Nuggetty Rd	5475 2472
Palm House B&B	Accommodation	2-6 High St	5475 2532
Terraces Miners Cottage	Accommodation	11 Fountain Street	9818 8512
Tressiders Cottage	Accommodation	54 High St	5426 3080
Tressiders Cottage	Accommodation	54 High St	5426 3080
Wywurri B&B	Accommodation	3 Templeton Street	5475 2794
Travelscene Castlemaine	Accommodation	175 Barker Street	5472 3822
Leckie Gallery	Art	9 Main Street	5475 1752
Maldon Gallery	Art	22-24 Main St	5475 2595
Maldon Marbles	Art	48 High Street	5475 1536
Carman's Tunnel	Attractions Specialities	Parkins Reef Road	5475 2667
Chocolade	Attractions Specialities	39 Main St	0418 524 945
Maldon Motor Museum	Attractions Specialities	46 High Street	5475 2222
Maldon Museum/Archives	Attractions Specialities	Shire Gardens Cnr High & Fountain Sts	5475 2041
Victorian Goldfields Railway	Attractions Specialities	Maldon Station Hornsby St	5476 4393
Maldon Golf Club	Attractions Specialities	Golf Links Road	5475 2005
Teddy and Me	Attractions Specialities	67 High St	5475 1228
Berrymans Cafe & Tea Rooms	Eateries	30 Main St	5475 2904
Cafe Maldon	Eateries	52 Main St	5475 2327
Calder House & Ruby's Rest.	Eateries	44 High St	5475 2912
Grand Hotel	Eateries	26 High Street	5475 2233
Maldon Hotel	Eateries	58 Main Street	5475 2231
Penny School Gallery	Eateries	11 Church Street	5475 1911
Stanyers Pottery - Studio	Eateries	Davies Lane	5475 2654
J & A Fitzpatrick - Butchery	Fooderies	19 Main St	5475 2271
Maldon Heritage Winery	Winery	Whitlocks Road	5473 4215
Nuggetty Vineyard	Winery	Maldon Sherbourne Road	5475 1347

Malmsbury

Hopewell Cottage	Accommodation	19 Ross Street	5423 2470
Malmsbury Hotel / Motel	Accommodation	85 Mollison Street	5423 2322
The Mill at Malmsbury	Accommodation	Calder Highway	5423 2267
Tin Shed Arts	Art	75 Mollison Street	5423 2144
Woodbine	Art	2644 Daylesford Rd	5423 2065
Malmsbury Bakery	Eateries	Mollison St	5423 2369
Malmsbury Hotel Motel	Eateries	85 Mollison St	5423 2322
The Mill at Malmsbury	Eateries	Calder Hwy	5423 2267
Malmsbury Bakery	Fooderies	Mollison St	5423 2369
Basalt Estate Winery	Winery	199 Zig Zag Road	5423 9108
Tarrangower Estate	Winery	Baldry Street	5423 2088

Mandurang

Lynnevale Estate	Accommodation	83 Cahills Road	5439 3635
Chateau Dore	Winery	Sutton Grange Road	5439 5278
Mandurang Valley Winery	Winery	77 Fadersons Lane	5439 5367
Tannery Lane Vineyard	Winery	Tannery Lane	5439 3227

Metcalfe

Morris' Country B&B	Accommodation	159 Kyneton - Metcalfe Road	5423 2435
Blackgum Estate	Winery	166 Malmsbury Road	5423 2933
Coliban Valley Vineyard	Winery	Redesdale Rd	0417 312 098

Moliagul

Mount Moliagul	Winery	Clay Gully Rd	0427 221 641

Mount Macedon

Blue Ridge Inn	Accommodation	1 Falls Road	5427 0220
Braeside	Accommodation	47 Taylors Road	5426 1762
Clonandra	Accommodation	553 Barringo Rd	5426 3453
Croft Cottage	Accommodation	4 Cedar Lane	5426 1592

THE HOUSE WHISPERER

FOR THE LACK OF A NAIL, THE STRUCTURE WAS LOST

Too often I have been a passerby to some old barn or house and over the years I've watched its derelict decline into a sad and sorry pile of pieces, knowing it need not have happened

Maybe you don't know one end of a hammer from the other.
Maybe you have some skills but don't know where to begin.
Maybe you like the angle its at and just want to keep it that way.
Maybe, until now, you haven't given it a thought.

My name is Frank Veldze and I am a specialist in almost everything to do with old buildings.
I may be able to help slow or reverse the aging process

Feel free to call me or send a photo of your project via email or Pixt.

I can be contacted via mobile: 0414 615205. home: (03)54232430.
or email: frankveldze@optusnet.com.au

MAKING A NATION FEDERATION EXHIBITION

A century later, the Making a Nation Federation Exhibition celebrates Bendigo's role in our Federal story, and highlights the city's dreams and hopes beyond 1901. This permanent exhibition sits within the former historic post office building where for the first time you can explore this fascinating heritage building and even send your own Morse Code messages.
Entry by gold coin donation.

Old Post Office, 51-67 Pall Mall, Bendigo
T: [03] 5444 4445 F: [03] 5444 4447
E: tourism@bendigo.vic.gov.au
www.makinganation.com

THE ALTAR BAR AT THE CONVENT GALLERY

Stylish and contemporary, The Altar Bar at The Convent Gallery serves a wide selection of light refreshments, cocktails and local wines. Sit by the fire, savour the magnificent views and relax in our little slice of heaven.

Daly Street. Daylesford 3460
T: [03] 5348 3211 F: [03] 5348 3339
E: visit@theconvent.com.au
www.theconvent.com.au

GREYSTANES MANOR

Greystanes Manor offers luxury boutique accommodation in the heart of Bendigo. Be pampered with a visit to our in-house beauty therapist and masseur, or stroll to nearby shops, restaurants and theatres.

Greystanes also offers excellent facilities for meetings and conferences, restaurant and private dining, and the most stunning backdrop for your wedding day. Greystanes Manor – the perfect setting for any occasion

57 Queen Street, Bendigo
T: (03) 5442 2466
www.greystanesmanor.com.au

AUTHOR

JOSEPH KINSELA

Joseph Kinsela is a historian, writer, professional musician, landscape gardener, and architectural historian with a Masters in Heritage Conservation from the University of Sydney.

He was an Anglican monk for twenty years, before leaving to pursue his music. His wide interests and expertise (including early years as a surveyor) mean that he sees place within patterns of land use, geology, social, economic and political history. His time as a monk and role as performer (with the Australian Opera and as a soloist on the United Kingdom oratorio circuit) give him a particular focus on the use of grand buildings.

Fascinated by the gothic revival in Australia, he is an expert on the work of Edmund Blacket. He has written guide books for Blacket's St Andrew's Cathedral Sydney, Christ Church St. Laurence Sydney, St Saviour's Cathedral Goulburn and St. Andrew's Braidwood. He now lives at Mt Macedon.

JOSEPH KINSELA'S ACKNOWLEDGMENTS

David Ryrie was my companion, sharing scenic and gustatory delights and the rigours of winter in the highlands. Without his car and company much less would have been achieved. Judith Pugh's project management and editorial discipline meant I stopped writing. Both leant a critical ear to my fascination with this captivating parcel of Australia.

PHOTOGRAPHY CREDITS

GARY CHAPMAN
(CHAPMAN'S FINE PHOTOGRAPHY)

Gary Chapman has lived in Central Victoria all his life and has been a professional photographer for 17 years. His day-to-day business includes weddings and commercial work, but he also specialises in large format panoramic landscapes and seascapes. Wherever he is, he makes time to tour the back roads searching for his next subject and that 'magical light'. To see more of Gary's stunning panoramas visit his website at or call in to his business for a friendly chat. Chapman's Fine Photography, 135 Mostyn St, Castlemaine.
Website: www.panoramaplus.com.au. Tel/fax 5470 5302

DARYL CHIBNALL (AEROVISION)

Daryl is a professional photographer and commercial pilot based in Ballarat, where he specialises in the discipline of aerial photography including GIS mapping and aerial survey. He owns and operates two purpose modified aircraft for capturing oblique and vertical aerial aspects using both medium and large format cameras. In 2000, after 28 years in graphic design, Daryl established AeroVision to satisfy two of his life-long passions, photography and flying. **Contact: daryl@aerovision.net.au**

GREG 'ARJUNA' GOVINDA

Arjuna's main photographic interests include travel, documentary and photojournalism - aiming to depict the beauty of spirit in action. He is also drawn to the mystical in nature, including the micro worlds of flowers. Arjuna's journeys have taken him around the world as well as into some of the depths of human experience. When not photographing he finds peace and joy in meditation, gardening, Aikido, Jungian psychology and Maitreya Theosophy. **Website: www.eagleheart.com.au. Tel: 5348 1414**

THEODORE HALACAS

Born in 1981 Theodore Halacas has enjoyed the benefits of a rural Australian upbringing interlaced with the culture of his migrant parents. A student of the Photographic Imaging College in Hawthorn, Theo won a first year Award in the State Film and Photography Festival, which was followed by a highly successful exhibition of portraits and landscapes in his home town, Castlemaine. Theo has worked in association with a number of studios including Ellio Rulli's highly-regarded Studio House Photography. He currently shares his time between Sydney, Melbourne and Central Victoria. **Website: www.theodorehalacas.com.au. Tel: 0403 030 522**

GEOFF HOCKING

Geoff Hocking is a Bendigo-born artist who lives today in the goldfields city of Castlemaine. Geoff has published a significant collection of books on the subject of Australia's Colonial past, with particular interest in the goldfields. His photographs reflect his interest in this history - of an environment littered with the memories of a golden past. Geoff is a collector of imagery, ephemera, paintings, engravings, tins, signs and printed memories which underpin his love of, and the retelling of, a good Australian story. **Contact: casbooks@netcon.net.au**

JOE MORTELLITI

Joe Mortelliti's passion for photography began when he was given a darkroom developing kit for his 13th birthday. He was hooked from the first time he saw images coming up in a developing tray.... These days Joe shoots on a digital camera, but he remains a purist and does not manipulate the images on computer. Joe and his wife Marion are working on a book of photographs covering Victoria and they are currently exploring the hidden corners of the state. Selected images printed on archival silver halide photographic paper are available on their website. Contact Lifestyle Images of Australia.
Website: www.images-australia.com.au

KATHERINE SEPPINGS

Katherine Seppings, local and international artist, writer and photographer, came to live in Chewton in 1983. A member of the National Trust Photographic Committee, Katherine was attracted to the Goldfields for visual and historical interests. Publications include "Women of the Hills", "The Complete Australian Bushfire Book", and "Fireplaces for a Beautiful Home". After travelling for many years, living and working in London and New York, Katherine returned to the Heart of Victoria, to be mother of daughter Maya. Exhibitions include "Characters of Castlemaine", "Around the World", "Nepal", "Back Streets of the World", "New York Graffiti" and "Castlemaine at the Crossroads". **Contact: kseppings@castlemaine.net**

ALISON POULIOT

My photography is an attempt to share a wonderment in the natural environment. My background as an environmental scientist and my insatiable wanderlust have led me to many remote and unique locations across the globe. Every time I'm invited to enter and seek the resonance of a place, it is a singular privilege and challenge. I always try to proceed with awareness, respect and a good dash of daring. I can be contacted at app@netcon.net.au regarding sales and commissions.

SANDY SCHELTEMA

Sandy Scheltema is a photojournalist based in Trentham, Central Victoria. Her assignments have taken her to Africa, India, Asia, The Pacific, Canada, and North and South America. Her favorite place though, is Central Victoria, where she enjoys photographing the changing light of different seasons.
Contact: sandys@netcon.net.au

CHRIS KIRWAN

Chris Kirwan was born in England in 1948 and came to Australia in 1954. A trained teacher and counsellor, he is presently Senior Counsellor and Head of Student Services at La Trobe University, Bendigo. An amateur photographer, he has so far managed to complete one semester of Photojournalism at La Trobe Bendigo. He prefers photographing landscapes, particularly using late afternoon light and often seeks out historical images such as old buildings and farm machinery. His first exhibition entitled "Looking through to the present" was presented at the Bendigo Pottery during Easter 2000. **Tel: 5439 6494**

PETE WALSH

Pete Walsh's passion for photography and the outdoors crystalized during a solo mountain bike cycling journey through remote Australia in 1988. Self taught, Pete has since worked as a photographer, focussing primarily on the Australian landscape. A selection of Pete's images are available as prints and cards from his web site and retail outlets in the Spa Country region.
Website: www.imagesofvictoria.com.au

OTHER PHOTOGRAPHERS: Christine Ramsay, Martin Hurley

BestShot!

Best Shot! publishes books, calendars, posters and cards that celebrate the landscapes, buildings and people of regional Victoria. Best Shot! is for people who understand 'the spirit of place'.

- www.bestshot.com.au

- Coffee-table guidebooks

- Large 'art' calendars

- Regular calendars

- Cards and posters

- Framed photographs

For more information:

See: www.bestshot.com.au
Email: click@bestshot.com.au
Freecall: 1300 66 49 43

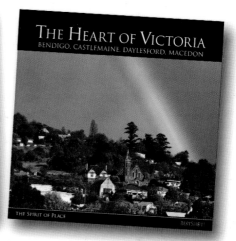